Climate change

A resource for students and teachers to support teaching about the nature of scientific enquiry and the strengths and limitations of scientific evidence

Written by Dorothy Warren
RSC School Teacher Fellow 1999–2000

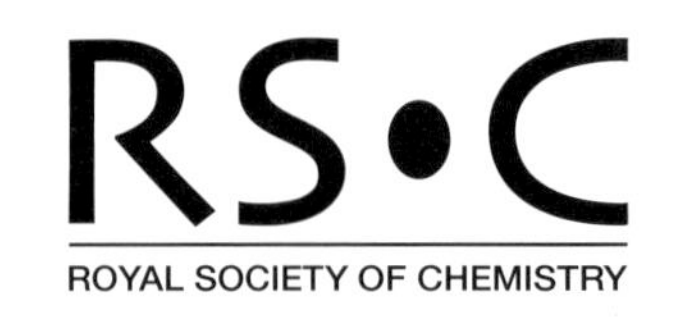

Climate change

Written by Dorothy Warren

Edited by Colin Osborne and Maria Pack

Designed by Imogen Bertin

Published and distributed by Royal Society of Chemistry

Printed by Royal Society of Chemistry

For further information on other educational activities undertaken by the Royal Society of Chemistry write to:

Education Department
Royal Society of Chemistry
Burlington house
Piccadilly
London W1J 0BA

Information on other Royal Society of Chemistry activities can be found on its websites:
http://www.rsc.org
http://www.chemsoc.org
http://www.chemsoc.org/LearnNet contains resources for teachers and students from around the world.

ISBN 0–85404–954–1

British Library Cataloguing in Publication Data.

A catalogue for this book is available from the British Library.

RS•C

Foreword

Too often young people think of science as a defined body of knowledge and scientists as people who know the answers. This resource attempts to present students with decisions to make, based on the data available to scientists from several different sources on climate change and global warming. It seeks to show that different trends can be observed over different timescales.

The resource also shows how, through the work of scientists such as Mario Molina, the political agenda can be changed for the good of mankind.

I hope you will use the resource with your students to bring them to a wider experience of how scientists work and the strengths and limitations of the scientific method and scientific evidence.

Professor Steven Ley CChem FRSC FRS
President, The Royal Society of Chemistry

RS•C

Using the resource on an Intranet

If your school or college has an Intranet you may wish to download the material in part or in whole to the Intranet to facilitate easy links for students. Alternatively you could use some of the web references given to design an interactive worksheet. Instructions for this are given below.

Designing an interactive worksheet
These instructions are for Microsoft Word.

- First you need to do some research on the Internet to find the relevant sites related to the topic and note down the website addresses, or use the sites already known and available from the resource.
- Choose an appropriate font, size and colour for your text.
- Type in the title of the page.
- Save this page as an HTML file in order to make the page 'live'.
- Type in the instructions and questions that you want the students to read.
- Type in the web page address (url). Always begin http://www (this will automatically make the page a hyperlink to the website you have typed and the text will turn blue.

Now you can make your work look more like a web page by placing lines, graphics, scrolling text and backgrounds in it.

- To place a horizontal line on your page, put the cursor where you would like the line to be, then click on insert, go to horizontal line, choose a line and click on OK.
- To add a background to your page, click on format, go to background, then fill effects option, choose an effect and click on OK.
- To add a picture to your work, place the cursor where you would like the picture to be, then click on insert, go to picture and clip art option. Choose a suitable picture for your work and click on OK. The picture size can be altered by moving the edges in or out or the picture can be moved to another place by dragging it over.
- To place scrolling text in your work, you need to highlight the words, click on insert, go to scrolling text, choose background colour, the speed of the scroll and press OK. (If you are going to print this work out, the scrolling text will not print out).

  Your work is now beginning to look like an interactive web page.
- When you are satisfied with the final product, click on file, go to web page preview and this will show you what your page looks like. (You cannot alter your page through the web page preview screen; you will have to go back to Word).
- Remember to do a spell check on your work, *ie* click on the ABC icon on the top of toolbar.
- Test out your page.

Now you are ready to use the page with students.

RS•C

Acknowledgements

The production of this book was only made possible because of the advice and assistance of a large number of people. To the following, and everyone who has been involved with this project, including the members of the science staff and students in trial schools, both the author and the Royal Society of Chemistry express their gratitude.

General
Colin Osbome, Education Manager, Schools & Colleges, Royal Society of Chemistry
Maria Pack, Assistant Education Manager, Schools & Colleges, Royal Society of Chemistry
Members of the Royal Society of Chemistry Committee for Schools and Colleges.
Members of University of York Science Education Group.
Jill Bancroft, Special educational needs project officer, CIEC, York
Donald Stewart, Dundee College, Dundee
Richard Warren, Mathematics Department, Ampleforth Colledge, York
Bob Campbell, Department of Educational Studies, University of York
Professor David Waddington, Department of Chemistry, University of York.
Jane Francis, Department of Earth Sciences, The University of Leeds
Stuart Monro, Our Dynamic Earth, Edinburgh
Barbara Knowles, National Environmental Research Council (NERC)
Jon Shanklin, British Antarctic Survey (BAS) (supplied data for ozone graphs)
Nicholas McWilliam, British Antarctic Survey (BAS) (supplied Antarctica map)
Rob MacKenzie, Institute of Environmental and Natural Sciences, University of Lancaster

Teachers & Trial Schools
David Billett, Ampleforth College, York.
Peter Bird, Alderwasley Hall School, Derbyshire.
Sandra Buchanan, Tobermory High School, Isle of Mull.
Louise Campbell, Greencroft School, Stanley.
Howard Campion, Fulford School, York.
Arthur Cheney, All Saints School, York.
Margaret Crilley, St Leonard's RC Comprehensive School, Durham.
Peter Dawson, Science adviser for York and the York Schools Secondary Science group.
John Davies, Hipperholme & Lightcliffe High School, Halifax.
John Edlin, Wolverhampton Grammar School, Wolverhampton.
Tim Gayler, Little Ilford School, London.
Carole Lowrie, Hummersknott School, Darlington
Greg McClarey, Blessed Edward Oldcorne RC High School, Worcester.
Lesley Stanbury, St Albans School, St Albans.
G. A. Thomas Ysgol Llanilltud Fawr, Vale of Glamorgan
Susan Vaughan, All Saints School, York.

The Royal Society of Chemistry would like to extend its gratitude to the Department of Educational Studies at the University of York for providing office and laboratory accommodation for this Fellowship and the Head Teacher and Governors of Fulford Comprehensive School, York for seconding Dorothy Warren to the Society's Education Department.

RS•C

Contents

RS•C

How to use this resource

At the start of the 21st century secondary education yet again underwent changes. These included the introduction of new curricula at all levels in England, Wales and Scotland and the Northern Ireland National Curriculum undergoing review. With more emphasis on cross curricula topics such as health, safety and risk, citizenship, education for sustainable development, key skills, literacy, numeracy and ICT, chemistry teachers must not only become more flexible and adaptable in their teaching approaches, but keep up to date with current scientific thinking. The major change to the science 11–16 curricula of England and Wales was the introduction of 'ideas and evidence in science', as part of Scientific Enquiry. This is similar to the 'developing informed attitudes' in the Scottish 5–14 Environmental studies, and is summarised in Figure 1.

In this series of resources, I have attempted to address the above challenges facing teachers, by providing:

- A wide range of teaching and learning activities, linking many of the cross-curricular themes to chemistry. Using a range of learning styles is an important teaching strategy because it ensures that no students are disadvantaged by always using approaches that do not suit them.
- Up-to-date background information for teachers on subjects such as global warming and Green Chemistry. In the world of climate change, air pollution and sustainable development resource material soon becomes dated as new data and scientific ideas emerge. To overcome this problem, these resource have been linked to relevant websites, making them only a click away from obtaining, for example, the latest UK ozone data or design of fuel cell.
- Resources to enable ideas and evidence in science to be taught within normal chemistry or science lessons. There is a need to combine experimental work with alternative strategies, if some of the concerns shown in Figure 1, such as social or political factors, are to be taught. This can be done for example, by looking at the way in which scientists past and present have carried out their work and how external factors such a political climate, war and public opinion, have impinged on it.
- Activities that will enhance student's investigative skills.

These activities are intended to make students think about how they carry out investigations and to encourage them to realise that science is not a black and white subject. The true nature of science is very creative, full of uncertainties and data interpretation can and does lead to controversy and sometimes public outcry. Some of the experiments and activities will be very familiar, but the context in which they are embedded provide opportunities for meeting other requirements of the curriculum. Other activities are original and will have to be tried out and carefully thought through before being used in the classroom. Student activities have been trialled in a wide range of schools and where appropriate, subsequently modified in response to the feedback received.

Dorothy Warren

RS•C

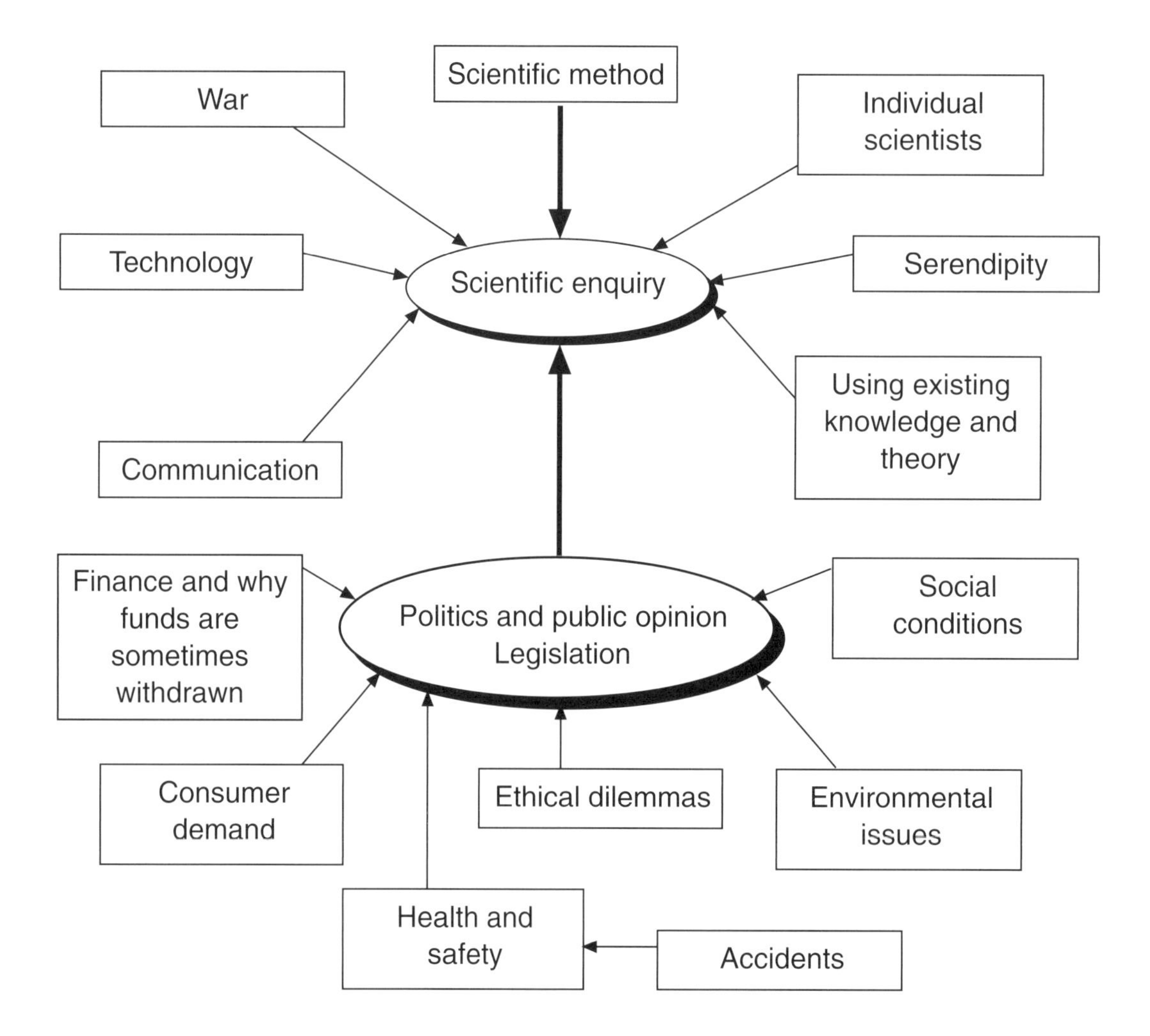

Figure 1 The factors influencing the nature of scientific ideas – scientific enquiry and the advancement of science

Maximising the potential use of this resource

It is hoped that this resource will be widely used in schools throughout the United Kingdom. However, as every teacher knows, difficulties can be experienced when using published material. No single worksheet can cater for the needs of every student in every class, let alone every student in every school. Therefore many teachers like to produce their own worksheets, tailored to meet the needs of their own students. It was not very surprising when feedback from trial schools requested differentiated worksheets to allow access to students of different abilities. In an attempt to address these issues and concerns, this publication allows the worksheet text and some diagrams to be modified. All the student worksheets can be downloaded in Word format, from the Internet via the LearnNet website, **http://www.chemsoc.org/networks/learnnet/ideas-evidence.htm** .This means that the teacher can take the basic concepts of the activity, and then adapt the worksheet to meet the needs of their own students. Towards the end of the teachers' notes for most activities there are some suggestions as to how the resource can be adapted to meet the needs of students of different abilities. There are also some examples of differentiated worksheets included in the resource.

RS•C

It is not envisaged that teachers will use every activity from each piece of work with an individual class, but rather pick and choose what is appropriate. For example some activities use high level concepts and are designed to stretch the most able student and should not be used with less able students *eg* Version 1 of the Mario Molina work.

Activities that involve researching for secondary information on the Internet contain hyperlinks to appropriate websites. To minimise the mechanical typing of the URLs and possible subsequent errors, the students can be given the worksheet in electronic form and asked to type in their answers. The websites are then only a click away.

Appropriate secondary information has been included in the teachers' notes for use in class when the Internet or ICT room is unavailable.

Unfortunately, from time to time website addresses do change. At the time of publication all the addresses were correct and the date that the site was last accessed is given in brackets. To minimise the frustration experienced when this happens, it is advisable to check the links before the lesson. If you find that a site has moved, please email both **LearnNet@rsc.org** and **education@rsc.org** giving full details so that the link can be updated on the worksheets on the web in the future.

Strategies for differentiated teaching

All students require differentiated teaching and it is not just an issue for those students with special educational needs. The following definition by Lewis[1] has been found to be quite useful.

'Differentiation is the process of adjusting teaching to meet the needs of individual students.'

Differentiation is a complex issue and is very hard to get right. It can be involved in every stage of the lesson i.e. during planning (differentiation by task), at the end of the activity (differentiation by outcome) and ongoing during the activity. Often teachers modify the activity during the lesson in response to feedback from the class. Differentiation does not only rely on appropriate curriculum material but is also concerned with maximizing learning. Student involvement and motivation effect the learning experience and should be considered and taken into account. It is therefore not surprising that differentiation is one of the areas of classroom teaching where teachers often feel under-confident. Most strategies for differentiated lessons are just applying good teaching practice eg varying the pace of the lesson, providing suitable resources and varying the amount and nature of teacher intervention and time.[2] Rather than just providing several examples of differentiated activities from the same worksheet, a list of strategies for differentiated teaching is presented, with some examples of how they can be used in the classroom. The examples can be found at the appropriate places in the text.

1. Using a range of teaching styles

A class is made up of different personalities, who probably have preferred learning styles. Using a range of teaching approaches makes it more likely that all students will be able to respond to the science that is being taught. The following examples have been included and can be found at the appropriate place in the resource.

Example – The versions of the Mario Molina worksheets
Version 2 is a more structured approach whereas Version 1 allows for discussion.

RS•C

2. Varying the method of presentation or recording
Giving the students some choice about how they do their work. There are many opportunities given throughout the resource.

3. Taking the pupil's ideas into account
Provide opportunities for students to contribute their own ideas to the lesson. For example when setting up an investigation allow different students the freedom to chose which variables they are going to investigate. The use of concept cartoons provides an ideal opportunity for students to discuss different scientific concepts. (see D. Warren, *The nature of science*, London: Royal Society of Chemistry, 2001.)

4. Preparing suitable questions in advance
Class discussions are important in motivation, exploring ideas, assessment etc. Having a list of questions of different levels prepared in advance can help to push the class.

5. Adjusting the level of scientific skills required
Example – Using symbol equations or word equations

6. Adjusting the level of linguistic skills required
Teachers may like to check the readability of their materials and of the texts they use. Guidance on this and on the readability of a range of current texts may be found at **http://www.timetabler.com/contents.html** (accessed June 2001).

7. Adjusting the level of demand on the student
Example – Mario Molina puts ozone on the political agenda Versions 1 and 2.

References

1. A. Lewis, *British Journal of Special Education*, 1992, **19**, 24–7.
2. S. Naylor, B. Keogh, *School Science Review*, 1995, **77(279)**, 106–110.

How scientists communicate their ideas

Effective communication is crucial to the advancement of science and technology. All around the globe there are groups of research scientists and engineers, in universities and in industry, working on similar scientific and technological projects. Communication between these groups not only gives the scientist new ideas for further investigations, but helps in the evaluation of data. Results from different groups will either help to confirm or reject a set of experimental data. Communication is vital when a company wants to sell a new product. Depending on the product the buyer will want to understand how it works and how to maintain it. Several of the employees will have to learn how to use the product, and respond quickly to changing technology and circumstances. Therefore the manufacturers must be able to communicate the science to prospective buyers.

Scientists communicate in a number of ways including:

- Publication in research journals
- Presenting papers at scientific conferences
- Poster presentations at conferences
- Book reviews by other scientists
- Publication on the Internet

RS•C

- Sales brochures
- Advertising flyers
- Television documentaries.

Publication in research journals
The article is written. The article must have an abstract, which is a short summary.

It is submitted to a journal.

The article is refereed by other scientists, working in a similar area. This is to check that the work is correct and original.

The article may be returned to the author to make changes.

The article is accepted and published by the journal.

The article is published.

Presenting papers at scientific conferences
Conference organisers invite scientists to speak on specific topics and projects.

An abstract is submitted to and accepted by the conference organisers.

The conference programme is organised and the speakers notified.

The scientist gives their talks, usually aided by slides or multimedia presentations which contain the main points.

There is usually time for questions after the talk.

The written paper is given to the conference organisers.

All the papers are published in the conference proceedings. This is usually a book.

Poster presentations at conferences
An abstract is submitted to and accepted by the conference organisers.

The conference programme is organised and the poster people notified.

During the poster session the authors stand by the posters, ready to answer any questions as the delegates read the posters.

Written papers may then be published in the conference proceedings.

Book reviews
Other scientists in the same field often review new books. The reviews are then published in scientific magazines and journals. The review offers a critical summary of the book. The idea of the review is to give possible readers an idea of the contents and whether it is suitable for the intended purpose.

Publishing on the internet
This is the easiest way to publish. Anyone can create their own web page and publish their own work. In this case the work is not refereed or checked by other people.

However, a lot of the information published on the Internet is linked to reputable organisations. In this case the articles will have been checked before they are published. Much of the information published on the Internet is targeted at the general public, and therefore the scientific ideas are presented in a comprehensible way. There are often chat pages so people can communicate their views and ask questions or request further information. The power of the Internet is that there is the opportunity to get immediate feedback to a comment or question.

Sales brochures

The information must be presented in an attractive and concise manner. After all you are trying to sell something. There should be a balance between technical information and operating instructions!

Advertising flyers

This must be written with the target audience in mind.

The information must be concise as there is limited space. The format must be attractive and should include pictures as well as writing. The flyer should also be quite cheap to produce.

Teaching students to communicate ideas in science

Students can be taught effective communication skills:

- By encouraging communication between students and a range of audiences in classrooms.
- By encouraging them to investigate like 'real scientists' by reporting their findings for checking and testing by others, and participating in two-way communication. (Communicating between groups, classes, partner schools, schools abroad perhaps via the Internet.)
- By setting investigations in a social context which offers the opportunity to communicate the project outside of the classroom. These work best when there is local interest.

When presenting investigative work to an audience, the student should consider the following:

- Who will be in the audience?
- What information does the audience need to know *eg* method, results and recommendations?
- How to present the information in an interesting and professional way *eg* should graphs be hand drawn or done on the computer?
- That the information offered convinces the audience that their investigation was valid and reliable.
- Poster presentations or display boards should be concise, since the space is limited.
- When speaking to audiences remain calm, speak clearly and slowly and try to be enthusiastic. Make sure that information on slides and OHTs can be read from the back of the room.

When writing a report of the findings of a scientific investigation for others to check and test, the emphasis should be on clarity. Another person is going to carry out the same investigation. The only information available is what is written in the report.

The report could be written under the following headings:

- Introduction
- Scientific knowledge
- Planning
- Table of results
- Graphs

RS•C

- Conclusions
- Evaluation
- Recommendations

Further background information

R. Feasy, J. Siraj-Blatchford, *Key Skills: Communication in Science*, Durham: The University of Durham / Tyneside TEC Limited, 1998.

Curriculum coverage

Curriculum links to activities in this resource are detailed at **http://www.chemsoc.org/networks/learnnet/climate.htm**

Curriculum links to activities in other resources in this series are detailed at

http://www.chemsoc.org/networks/learnnet/ideas-evidence.htm

Health and safety

All the activities in this book can be carried out safely in schools. The hazards have been identified and any risks from them reduced to insignificant levels by the adoption of suitable control measures. However, we also think it is worth explaining the strategies we have adopted to reduce the risks in this way.

Regulations made under the Health and Safety at Work etc Act 1974 require a risk assessment to be carried out before hazardous chemicals are used or made, or a hazardous procedure is carried out. Risk assessment is your employers responsibility. The task of assessing risk in particular situations may well be delegated by the employer to the head of science/chemistry, who will be expected to operate within the employer's guidelines. Following guidance from the Health and Safety Executive most education employers have adopted various nationally available texts as the basis for their model risk assessments. These commonly include the following:

Safeguards in the School Laboratory, 11th edition, ASE, 2001

Topics in Safety, 3rd Edition, ASE, 2001

Hazcards, CLEAPSS, 1998 (or 1995)

Laboratory Handbook, CLEAPSS, 1997

Safety in Science Education, DfEE, HMSO, 1996

Hazardous Chemicals. A manual for science education, SSERC, 1997 (paper).

Hazardous Chemicals. An interactive manual for science education, SSERC, 1998 (CD-ROM).

If your employer has adopted more than one of these publications, you should follow the guidance given there, subject only to a need to check and consider whether minor modification is needed to deal with the special situation in your class/school. We believe that all the activities in this book are compatible with the model risk assessments listed above. However, teacher must still verify that what is proposed does conform with any code of practice produced by their employer. You also need to consider your local circumstances. Is your fume cupboard reliable? Are your students reliable?

Risk assessment involves answering two questions:

- How likely is it that something will go wrong?
- How serious would it be if it did go wrong?

How likely it is that something will go wrong depends on who is doing it and what sort of training and experience they have had. In most of the publications listed above there are suggestions as to whether an activity should be a teacher demonstration only, or could be done by students of various ages. Your employer will probably expect you to follow this guidance.

Teachers tend to think of eye protection as the main control measure to prevent injury. In fact, personal protective equipment, such as goggles or safety spectacles, is meant to protect from the unexpected. If you expect a problem, more stringent controls are needed. A range of control measures may be adopted, the following being the most common. Use:

- a less hazardous (substitute) chemical;
- as small a quantity as possible;
- as low a concentration as possible;
- a fume cupboard; and
- safety screens (more than one is usually needed, to protect both teacher and students).

The importance of lower concentrations is not always appreciated, but the following table, showing the hazard classification of a range of common solutions, should make the point.

Ammonia (aqueous)	irritant if ≥ 3 M	corrosive if ≥ 6 M
Sodium hydroxide	irritant if ≥ 0.05 M	corrosive if ≥ 0.5 M
Ethanoic (acetic) acid	irritant if ≥ 1.5 M	corrosive if ≥ 4 M

Throughout this resource, we make frequent reference to the need to wear eye protection. Undoubtedly, chemical splash goggles, to the European Standard EN 166 3 give the best protection but students are often reluctant to wear goggles. Safety spectacles give less protection, but may be adequate if nothing which is classed as corrosive or toxic is in use. Reference to the above table will show, therefore, that if sodium hydroxide is in use, it should be more dilute than 0.5 M (M = mol dm^{-3}).

CLEAPSS Student Safety Sheets

In several of the student activities CLEAPSS student safety sheets are referred to and recommended for use in the activities. In other activities extracts from the CLEAPSS sheets have been reproduced with kind permission of Dr Peter Borrows, Director of the CLEAPSS School Science Service at Brunel University.

- Teachers should note the following points about the CLEAPSS student safety sheets:
- Extracts from more detailed student safety sheets have been reproduced.
- Only a few examples from a much longer series of sheets have been reproduced.
- The full series is only available to member or associate members of the CLEAPSS School Science Service.
- At the time of writing, every LEA in England, Wales and Northern Ireland (except Middlesbrough) is a member, hence all their schools are members, as are the vast

majority of independent schools, incorporated colleges and teacher training establishments and overseas establishments.

- Members should already have copies of the sheets in their schools.
- Members who cannot find their sheets and non-members interested in joining should contact the CLEAPSS School Science Service at Brunel University, Uxbridge, UB8 3PH; tel. 01895 251496; fax. 01895 814372; email science@cleapss.org.uk or visit the website **http://www.cleapss.org.uk** (accessed June 2001).
- In Scotland all education authorities, many independent schools, colleges and universities are members of the Scottish Schools Equipment Resource Centre (SSERC). Contact SSERC at St Mary's Building, 23 Holyrood Road, Edinburgh, EH8 8AE; tel: 0131 558 8180, fax: 0131 558 8191, email: sts@sserc.org.uk or visit the website **http://www.sserc.org.uk** (accessed June 2001).

RS•C

This page has been intentionally left blank.

RS•C

Introduction

Climate change is a very complex subject. Scientists from many countries around the world strive to understand how climate works and the try to prepare for the changes that may lie ahead. There is a wealth of information available at numerous websites offering a number of different opinions, often with a bias towards the interests of those who produced the resources. Students should be encouraged to look at information from a variety of sources and then try to make their own minds up based upon what they find out.

In the first section the students will investigate climate change, paying particular attention to data analysis and evaluation. The activities are designed to challenge some of the views possibly already held by the students and should provoke some interesting group discussions.

The ozone activity looks at the history of chlorofluorocarbons (CFCs) and the ozone hole and shows how scientists like Mario Molina can influence the behaviour of governments, the chemical industry and the general public. In both activities students are encouraged to use the Internet to find out further information.

RS•C

Global warming

Teachers' notes

Objectives

- To understand the greenhouse effect.
- To understand that global warming is a complex issue full of uncertainties and controversies. Science is not clear cut and it does have its limitations.
- To collect data from different sources, to consider the evidence and evaluate the evidence. To understand that data collected from different sources does not always lead to the same results.
- To give an introduction to computer simulated models of climate change.

Outline

This resource includes a number of different types of activities and worksheets, which can be used to promote group discussion.

- Introducing global warming and understanding the greenhouse effect.
- **A global warning** – using data from the past to predict the climate of the future.
- Methods of collecting climatic data.
- The work and lifestyle of scientists collecting climatic information and data in Antarctica.
- The Earth is getting warmer, or is it?
- Addressing the ethical issues arising from global warming. What should we do about it?

Teaching topics

This material is suitable for students in the 11–16 year old age range. With modifications it could form the basis of a post-16 environmental science unit. The activities can be included when teaching about changes to the Earth and atmosphere and the effect of burning fossil fuels on the environment. It is an excellent context for teaching about the nature of scientific data and how to handle it.

Background information

Global warming is a very large area of scientific uncertainty. There are literally thousands of scientists working around the world, trying to understand and formulating models that will predict the consequences of global warming. It is an extremely complex procedure and there are hundreds of variables that come into the model. No one knows if the models being used are accurate. What will the climate will be like in 50 or 100 years time? No one can know for certain. However, there are real fears that rapid climate change will have a dramatic impact on life on Earth. Over the last 10,000 years the Earth has experienced a very stable climate and life has adapted to it. Recently, however, the Earth has seen an increase in temperature change and many scientists now believe that there is a direct link between this warming and emissions of greenhouse gases such as carbon dioxide (CO_2) and nitrogen oxides (NO_x) caused by human activities. In the past the greenhouse gases were in balance. It is feared that, at the present rate of increase of greenhouse gases, the natural balance will be upset. It takes millions of years for fossil fuels to form but only a few minutes for them to burn, releasing large quantities of CO_2 in to the air. In the past fluctuations in CO_2 levels have been explained by natural causes such as volcanic eruptions and the number of phytoplankton in the sea. There are currently several other theories about global warming such as positive and negative feedback systems of ocean currents and the position of the Earth in space.

As a result of the warming, global sea levels are expected to rise by a further 15 to 95 cm by the year 2100 because seawater expands when heated and some glacial ice will melt. Extreme events such as heavy rains and droughts are expected. There may be more hurricanes, typhoons and cyclones. Most of the predictions are based on rather slow changes in temperature, but when you look back through history there has been very sudden temperature changes which have had dramatic effects to life. The Atlantic Ocean conveyor belt system is one of the present theories that is under investigation that could explain these rapid changes in temperature. At the moment the conveyor belt (commonly known as the Gulf Stream) is switched on, thus keeping Britain and Northern Europe warm. If the conveyor was to be switched off, the story would be very different and without the Gulf Stream, there would be rapid cooling and we would be plunged into the next ice age. Some scientists believe that the conveyor belt could be switched off if large amounts of fresh water were to enter the Atlantic Ocean at the critical point. This fresh water could come from continuous rainfall as a result of a warmer, wetter climate and/or rapid melting of polar ice. Detailed data collected from ice cores have shown that in the past some temperature changes have occurred very quickly (within a hundred years). The main worry of some scientists is that the ever-increasing emissions of greenhouse gases that are causing a rapid increase in global temperature may result in the Atlantic Ocean conveyer system being switched off.

The real problem for the scientist is that climate change is very difficult to predict.

Using oxygen isotopes as temperature indicators

This method is commonly used to indicate temperatures using seabed sediments and ice cores. Students will need to understand the term isotopes and know the difference between oxygen-16 ($^{16}_{8}O$) and oxygen-18 ($^{18}_{8}O$).

Fossil shells are often made up of calcium carbonate. During its lifetime, the creature that inhabited the shell slowly extracted oxygen from water to build up its shell. Experiments have shown that the proportion of oxygen-16 and oxygen-18 in the shell is sensitive to the temperature. This is the temperature of the water the creature lived in. So scientists can estimate past ocean temperatures by determining the oxygen isotope ratios in the shell. Both surface temperatures and seabed temperatures can be determined by analysing the fossils of creatures that lived near the surface and on the bottom of the sea.

At high temperatures more of the lighter isotopes escape, thus a high $^{18}O : {}^{16}O$ ratio indicates a higher temperature that a low ratio does. The idea is simple, but in practice it is quite difficult to determine the individual layers of sediment.

Atmospheric temperatures can be measured by determining the ratio of oxygen isotopes in the layers of an ice core. Each year a new distinct layer of ice is formed, so it is easy to determine the age of the ice.

Using leaves as thermometers

Plants adapt to their environment. For example leaves which evolve in shaded, humid conditions are large and thin. If the same type of leaf is exposed to high levels of light, then it evolves to become small and thick with large cuticles. The conflicting demands of water conservation, gas exchange and light capture are governed by the laws of physics (evaporation and gas diffusion). Plants solve these dilemmas by physically adapting to meet the needs of the immediate environment. Within certain limits these adaptations are so constant that climatic comparisons can be made.

In 1915, two American botanists I.W. Bailey and E.W. Sinnott noted that leaves of modern woody 'broadleaved' flowering plants, *eg* fig and willow, tend to have smooth margins in warm climates but toothed 'jagged' ones in cooler climates. In the late 1970s, Jack Wolf, another American, by using modern species growing in drought free environments of south east Asia, was able show that there was a linear relationship between the mean annual temperature in degrees centigrade (°C) and the percentage of entire margined leaf species. The relationship breaks down if there is a limited water supply.

Today this method is used for estimating the climate millions of years ago.[1,2] Analysis of fossilised leaves indicate the temperature. This method is however subject to large sources of error and a large volume of data is required to get meaningful results.

Sources of information

- For background information on global warming, including a history of scientific discoveries since 1896, myths and facts about global warming / greenhouse effect and initiatives set up by a variety of different countries, visit the global warming website that will link you to 10 other relevant sites.
  http://www.ulisse.it/~irrsaege/globaw/siti_ing.html (accessed June 2001)
- The United Nations Environment Programme website is a place to visit if you are interested in what the politicians are thinking.
  http://www.unep.ch/ (accessed June 2001)
- You may like to get involved in the Science Across the World project on the Greenhouse effect. Go to **http://www.scienceacross.org** (accessed June 2001) click on useful links and then choose global warming.
- The Casino-21 experiment, a climate simulation of the 21st century. For the experiment to work millions of people are needed to run a simulation on their computer. The idea is to predict the climate in 2050. Details of the project and how to register can be found at
  http://www.climate-dynamics.rl.ac.uk/~hansen/casino21.html (accessed June 2001).
- **http://www.exxon.mobil.com** (accessed June 2001)
- **http://www.bp.com/default.asp** accessed June 2001)
- G. Best, *Environmental Pollution Studies*, Liverpool: Liverpool University Press, 1999.
- S. Lamb and D. Sington, *Earth Story, The shaping of our world*, London: BBC Worldwide Ltd., 1998.
- *Climate Change, scientific certainties and uncertainties*, NERC, Polaris House, North Star Avenue, Swindon, SN1 1EU. Email requests@nerc.ac.uk (This leaflet also gives a list of climate change contacts.)

RS•C

- C. O'Driscoll, *Chemistry in Britain*, 2000, **36(2)**, 28. (**http://www.chemsoc.org/chembytes/ezine/2000/toolkit_feb00.htm**) (accessed June 2001)
- A. Jones, *Chemistry in Britain*, 2000, **36(2)**, 33. (**http://www.chemsoc.org/chembytes/ezine/2000/toolkit_feb00.htm**) (accessed June 2001)

RS•C

A global warning

Teaching tips

The student sheet, **A Global Warning - is the Earth getting warmer?** is intended to introduce the theme of global warming as a controversial scientific issue. Using an OHT of the sheet to initiate a class discussion, find out what the class already know about global warming and if they have any strong views on the subject. For example, do they see global warming as a controversial subject or, in their minds, is it all clear cut? Do they understand what global warming and the greenhouse effect mean? What are the problems associated with global warming? Why are so many governments concerned with it?

The sheet has been presented as a conversation between two young children. This is to try and get over the point that even though most people think that they understand global warming, the fact is that it is an extremely complex subject and many scientists disagree. The sheet aims to challenge existing ideas about global warming. This may cause cognitive conflict. The students should be encouraged to discuss their ideas before they go on to look at the data.

The newspaper headlines have been deliberately chosen to create an air of uncertainty relating to the scientific world and to introduce some of the vocabulary. For example, some students may wonder why there might be another ice age if the world is warming up. Through discussion with the class, highlight some of the scientific problems that face scientists *eg* data collection when there are so many variables, or the use of models to predict future temperatures and levels of greenhouse gases. Stress that all these factors and many more lead to scientists being uncertain about the future of global warming.

The purpose of the introduction is to highlight some of the areas of uncertainty and not to try and answer all the questions. After looking at climate data there will an opportunity for the students to formulate their own opinions on the subject and to debate any future steps that should be taken by scientists and governments.

For this activity to be successful it must be introduced through discussion otherwise the students will not appreciate the complexity of the problem, and they will not see the point of interpreting different types of data.

Key words from the newspaper articles that may need introducing:

- Greenhouse effect;
- Global warming;
- Modelling.

The greenhouse effect and global warming This sheet presents students with an opportunity to explore the greenhouse effect and global warming in more detail, through a 'cut and stick' approach. It should be used before going on to look at the data, if the class are unfamiliar with the greenhouse effect.

Climate models are computer programmes that are used to try and reproduce climatic data by feeding in lots of different variables such as wind speed and direction, rainfall, temperature, concentration of greenhouse gases etc. When the scientists believe that the model gives a good reproduction of what actually happens, they can use it to predict the future. For example, by feeding in different concentrations of greenhouse gases they can predict by how much the earth surface temperature will rise. This information could then be used by international governments to set future greenhouse gas targets. (See note on Casino-21, page 4.)

Interpreting climatic data

This phase of the lesson involves the interpretation of real climatic data. The data is presented in the form of temperature – time graphs in the **Looking at the data** worksheets. Each graph is followed up by specific questions. Questions 1–20 are suitable to be used as individual class work or homework, as each question follows on from the last.

The International Panel of Climate Change (IPCC) temperature time data has been calculated and presented relative to the 1961–1990 average. When you look at the graphs, you will see that this average works out at zero. It is important to stress that it is not necessary to present absolute temperatures, because we are interested in looking at the temperature differences. The data file is found at **http://www.cru.uea.ac.uk/cru/data** (accessed June 2001).

Finally students should compare and discuss their graphs predicting the temperature change over the next twenty years.

The extension question could be discussed in groups. It may cause some difficulty as it is really intended to act as a thought provoker for the next session. There are two key ideas to stress here:

- In order to predict future temperature changes, a full picture of the past is required.
- Data collected by different methods can be very different, so where possible, scientists try to use data from more than one source. Which data set is the most accurate or reliable?

Acknowledgements for student worksheets

Newspaper headlines – © Times Newspaper Limited, (27/7/1989, 11/9/1994, 27/9/1995, 13/8/1998, 6/1/1999).

Data source for questions 1–10, Intergovernmental Panel on Climate Change.

Data source for questions 11–22, Central England Temperature data - Crown copyright, The Meteorological Office, Hadley Centre for Climate Prediction and Research.

Resources

- OHP
- Student worksheets:
  – A Global Warning – is the Earth getting warmer? – OHT Master for introductory discussion
  – The greenhouse effect and global warming
  – Looking at the data – temperature changes over the last century
  – Looking at the data – temperature changes over several centuries
  – Looking at the data – extension sheet

Timing

Allow 1 hour for basic coverage. Some classes will need longer if they do **The greenhouse effect and global warming** sheet or if they get very involved in the discussions.

RS•C

Adapting materials

Some teachers may feel that the questions are repetitive, in which case they could reduce the number of graphs that are analysed. However, care must be taken not to distort the overall aims of the material.

Methods of collecting temperature data

In this section teachers are not expected to cover all the material, rather choose activities that are suitable for each class.

Introduce this lesson by discussing the answer to the extension question in the previous session. Try to get as many ideas as possible before introducing the methods that are actually used by scientists gives factual information about how ice and sediment cores are drilled in the field and later analysed.

Resources

- Student worksheets:
  - Information sheet – collecting climatic data from ice and sediment cores
  - Collecting climatic data that is millions of years old.

Looking at real temperature and time data taken from North Atlantic sediment cores and ice cores in Greenland and Antarctica

This could be introduced by reminding the class that during the 20th century the Earth's average temperature has increased by approximately 0.5 °C. However, it is difficult to predict what will happen in the future when you are looking at a small time period, because every so often there are temperature fluctuations, such as in the 18th century, just after the little ice age.

In this lesson the students analyse temperature data going back 220,000 years and then look at the results of some temperature models that go back millions of years. By looking at levels of greenhouse gases, the students are asked to try and find patterns and links between the level of greenhouse gases and the Earth's temperature. This leads naturally into the final section, namely, man's influence on global warming. The question is posed 'what should we be doing about global warming?'

Resources

- Student worksheet:
  - Looking at climatic data from the past

Life as a scientist working in Antarctica

This section aims to give an overview of life working in Antarctica.

The student worksheets include an interview with Dr Jane Francis of Leeds University who has been part of past Antarctic expeditions. Some of the leaflets and videos listed in the resources section can also be used. The task at the end asks students to write an advert for expedition scientists. The advert should focus on the type of person required, the qualifications needed and should include something about the scientific work. This could be used in careers sessions.

Resources

- Student worksheet:
  – Life as a scientist in Antarctica
- Internet access
- The British Antarctic Survey, BAS, website **http://www.Antarctica.ac.uk** (accessed June 2001). Contact Schools Liaison Officer, British Antarctic Survey, High Cross, Madingley Road, Cambridge CB3 0ET, UK. Telephone 01223-221400. Enquiries from schools and students: schools@bas.ac.uk
- The old BAS website still has information for schools **http://www.nerc-bas.ac.uk/nerc-bas.html** (accessed June 2001)
- Antarctica The White Laboratory – a 25 minute video available from BAS
- Living & Working in Antarctica – an information leaflet available from BAS
- US Antarctic Resource Centre **http://usarc.usgs.gov/** (accessed June 2001)
- **http://www.antarcticanz.govt.nz/Pages/InfoEducation/Education.msa** (accessed June 2001).

Timing

It is suggested that no more than 1 hour is spent on this section.

Opportunities for ICT

- Internet based research
- Writing a job advert.

Opportunities for other key skills

- Communication – writing a job advert
- Application of number – working out percentages and using the results to determine temperatures from graphs
- Working in groups.

RS•C

The Earth is getting warmer

Teaching tips

This activity should be introduced by a class discussion based on the information on the student sheet, which can be used as an OHT.

The use of this section is threefold:

- It presents an opportunity to recap on what has been achieved so far and to make sure that everyone is following the argument.
- It introduces the possible causes of global warming; supporting the greenhouse gas theory with some real data.
- It asks the questions 'what will the effect of global warming be?' and 'should we be doing anything about it?'

The aim of the final section of speech bubbles is to introduce the ideas about what will happen when the world warms up and what should we be doing about it? It provides a further focus for a class or group discussion.

The future climate activities page offers a choice of different activities.

It is not intended that all of the activities should be done by everyone. Some teachers may give a free choice, whereas others might decide that only some are appropriate for a particular class. You may wish to give the students some help by supplying them with more information.

Other greenhouse gases
The aim of this activity is to show that carbon dioxide is not the only greenhouse gas and the recent increase in global temperatures may be attributed to an increase in the amounts of CO_2, CH_4, CFCs and O_3 in the atmosphere due to human activities.

Resources

- Student worksheets:
  - A Global Warning – the Earth **is** getting warmer
  - Looking at the data – the Earth is getting warmer
  - Other greenhouse gases
  - Theories about global warming
  - Future climates – activity page
- Internet access

Timing

Allow 30 minutes for the discussion section. The other activities are research based and are good for homework and/or another 1 hour lesson.

Adapting materials

Some teachers may wish not wish to offer all the activities on the **Future climates – activity page** and therefore wish to delete some of them.

Opportunities for using ICT

- Internet research
- Modelling
- Writing a report, which may include downloaded photographs or graphs.

Opportunities for other key skills

- Communication – designing a questionnaire and letter writing
- Working in groups.

RS•C

Answers

The greenhouse effect and global warming

1. –18 °C
2. The surface temperature of the earth is slowly warming up.
3. Coal fired power stations, gas/ oil central heating, any type of fires, vehicles etc.

Looking at the data – temperature changes over the last century

1. The general trend shows an increase in temperature. In the early 80s the temperature was stable for about three years before it started to increase again.
2. The maximum temperature change is about 0.25 °C.
3. Most students will sketch a graph which shows another increase over the next 20 years. Some students will copy the exact shape of the curve *ie* showing a plateau around 2002–2005. Accept a bigger increase – as some students may already know the connection between CO_2 level and temperature.
4. In the period 1960–1980 the overall temperature change was approximately zero. There was a steady decline in temperature (approximately 0.2 °C) between 1960 and 1975, when it started to increase.

   Whereas in the period 1940–1960, the temperature fluctuated raising to a maximum of about 0.05 °C and a minimum of about –0.05 °C.
5. Some students may chose to keep their 2000–2020 graph the same (based on previous knowledge). Other students may redraw their graph showing a steady decline in temperature by about 0.2 °C or they might show the graph fluctuating.
6. 1920–1940 saw a steady temperature increase from approximately –0.35 to + 0.00 °C, whereas between 1900–20 a decline in temperature was seen in the first decade followed by a steady increase in the second decade. The overall temperature change was approximately –0.2 °C.
7. Accept any reasonable graph.
8. The overall temperature increase in the 20th century is approximately 0.5 °C.

9&10

It should be pointed out that there is no correct answer because so many variables are involved. This data alone does not give enough evidence to make an accurate prediction of what will happen, but the overall temperature trend over the century has been to increase, so this might favour a graph showing an increase in temperature.

11. Yes the overall trend is the same, but there are some differences in the smaller temperature fluctuations.
12. Approximately the same +0.5 °C.
13. Data has been collected using different instruments, at different times of the day and night, in different places *ie* a fair test has not been carried out. Different variables are taken into account at different times.

RS•C

Looking at the data – temperature changes over several centuries

1. Yes, it was cooler, the temperature change oscillated more.

2. 19th century.

3. Approximately 1699.

4. It took about 4 years to warm up by a degree.

5. Temperatures reach a similar maximum temperature and then in the late 1730s the temperature fell again.

6. Similar temperatures.

7. Again, stress that the 'correct' answer is unknown, accept any reasonable attempt.

8, 9, 10. Teachers may wish each group to report back.

Looking at the data – extension sheet

At this point it may be appropriate to summarise the results so far.

1. Conclusion – the evidence considered so far indicates that the surface temperature of the Earth has increased by about 0.5 °C during the 20th century, but the evidence is too limited to make firm predictions about future temperature changes.

2. In practice several different methods are used. Geologists look for fossilised / petrified plants and animals, in sediment cores from fresh water lakes and the seabed. Surface rocks can also be used to reveal clues. They use the temperatures that these animals lived in to determine the temperature. Ice cores are used to determine the temperature by looking at the ratio of oxygen isotopes. Air pockets are sampled to reveal CO_2 levels from past climates. Accept any reasonable answers.

Collecting climatic data that is millions of years old

1. The main problem is that the ecology may have evolved and the plants are different. The plants growing in the Southern hemisphere are not as well documented as the plants in the Northern hemisphere.

2. **a)** 40% smooth = 12 °C
b) 70% smooth = 22 °C
c) 20% smooth = 7 °C
d) 90% smooth = 28 °C
e) 30% smooth = 10 °C
f) 50% smooth = 16 °C

3. **a)** Photograph A, the tree rings are wide indicating a lot of growth during the season. This tree grew very well under a greenhouse climate with plenty of water and warm temperatures. It is 100 million years old.

b) Photograph B, some of the rings are closer together than others, indicating years when the tree did not grow much, due to lack of water.

c) Photograph B shows the occasional drought, most of the time it grew well. Note the round holes were formed by boring worms when the tree was driftwood. The holes were then filled with sea floor sediment. The wood is now petrified.

Extension question – The answer should indicate some understanding that modern instruments are much more sensitive than they were even fifty years ago. Also the readings are now taken from all over the world, whilst early readings are limited geographically. Temperatures obtained from oxygen isotopes are measured using very sensitive equipment, but the method is built upon several assumptions. Temperatures collected by fossil methods are much less accurate because it is much harder to get large enough sample sizes and some temperatures are estimated using the nearest living relative! Exact measurements are not made.

Looking at climatic data from the past

1. There has been a small temperature decrease before stabilising out.

2. 16,000 and 50,000 years ago.

3. 12000, 22000, 36000 years ago.

4. The last time the temperature was similar to that today, it was followed by a gradual decrease over the next 10000 years. This could happen again.

5. It has not really varied but been stable.

6. Approx. 124–128000 years ago.

7. 5 °C lower than today.

8. 20–30000, 60–70000 and 140–150000 years ago.

9. 140–150000 years ago.

10. The temperature rose to reach the maximum recorded temperature (greater than today's temperature). Over the next 200 years the temperature rapidly dropped to today's temperature, where is stabilised out over the next 400 years before it continued to drop further.

11. Temperatures continued to drop until they reached a minimum.

12. It may follow the pattern of 130000–100000 years ago, plunging us into the next ice age.

13. Last 10000 years have seen a stable temperature.

14. Sediment cores show temperature changes of 10 °C, whereas the ice cores show changes of only 5 °C.

15. Yes, they follow the same general pattern.

16. Different methods of collection. Maybe harder to determine the timescale in sediments as they take longer to form.

17. Problems – Which set of data is correct? Should we only look at patterns and not actual numbers? Question – Why should the temperature vary so much? What causes the temperature trend to change direction?

18. **a)** Warmer that today.
b) 3
c) One every 100–200 million years.

19. Not really, but it does show that the Earth's climate has changed many times in the past and will change again.

Looking at the data – the Earth is getting warmer

1. The evidence strongly suggests that there is a firm link between the amount of CO_2 in the air and the average surface temperature of the Earth.

Other greenhouse gases

1. It increases.
2. Yes, because the data appears to show a link between the temperature and the amount of CH_4 in the atmosphere.
3. 1800
4. It increased at a much faster rate.
5. 1950s
6. Around 1950
7. Yes, because the global temperature increase since 1950 corresponds to the increased levels of these gases.
8. CH_4 at 10 years
9. Accept N_2O or CO_2
10. If time permits teachers could give the opportunity for some pupils to present their speeches to the class. In which case, please specify how long the speech should be. You may need to make some of the other worksheets such as **The greenhouse effect and global warming** available. This could be carried out as a group activity.

A Global Warning

Is the Earth getting warmer?

The greenhouse effect and global warming

A garden greenhouse keeps plants warmer than they would be outside. It does this because the glass traps some of the Sun's radiation energy. The atmosphere keeps the Earth warm in a similar way. Without the greenhouse effect the earth would be about 33 °C cooler than today's pleasant average of 15 °C. Greenhouse gases include carbon dioxide, oxides of nitrogen, methane, chlorofluorocarbons (CFCs) and water.

Show that you understand the greenhouse effect by cutting out the labels and sticking them on to the diagram below.

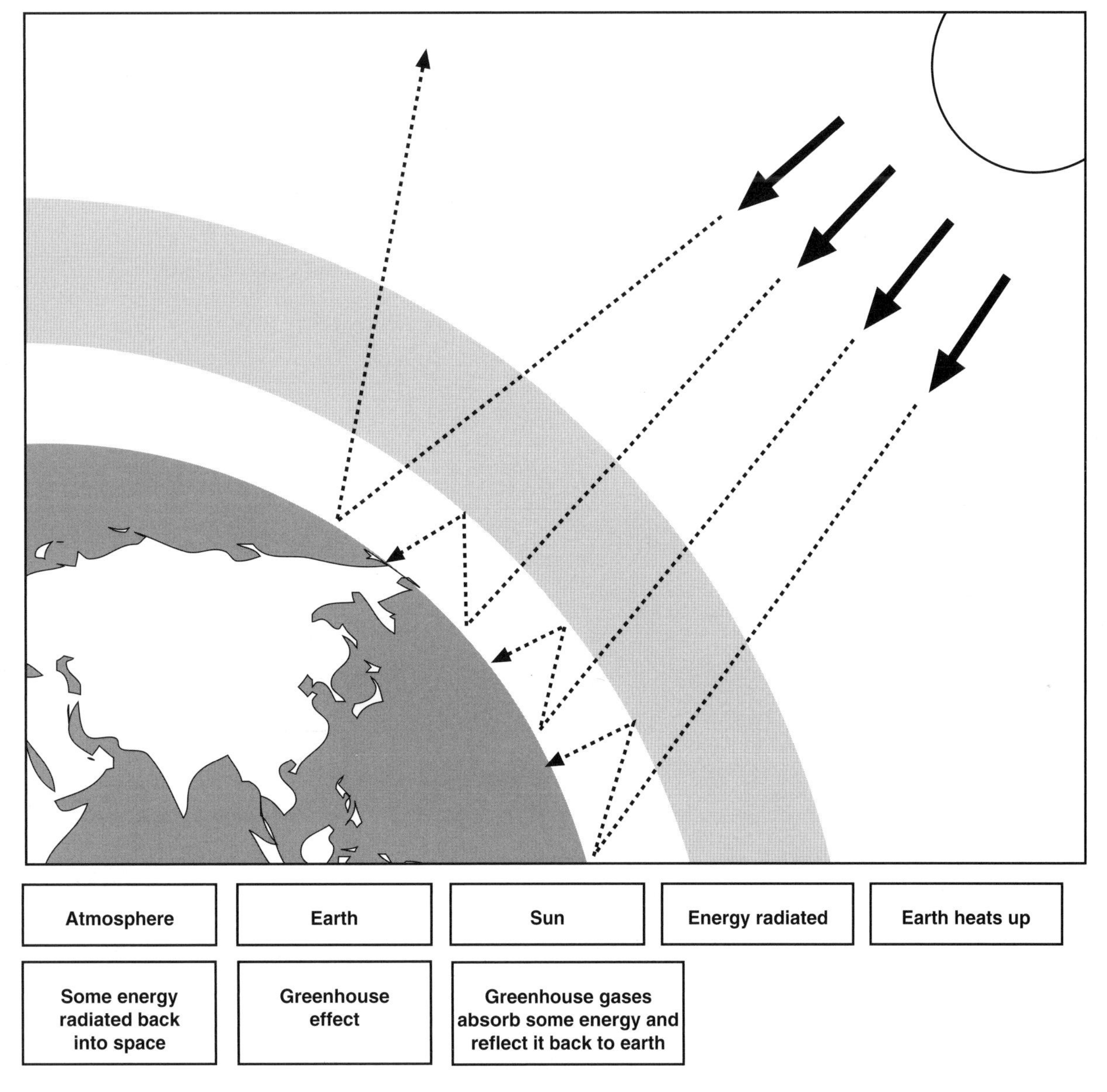

Questions

1. If the greenhouse effect did not exist, what would the normal temperature of the earth be?
2. What do you think is meant by global warming?
3. List all the things you can think of that give off greenhouse gases – *eg* cars.

Looking at the data – temperature changes over the last century

Study the graphs, published by the International Panel of Climate Change (IPCC) and then answer the questions. These are temperatures above and below the average for the period 1961–1990.

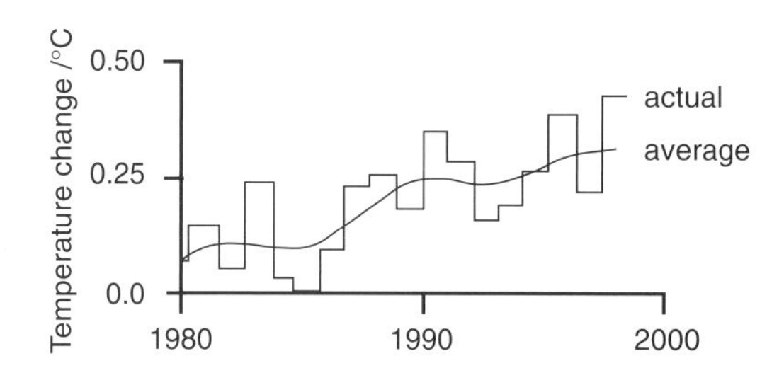

1980–2000 data

1. Describe how the temperature of the earth has changed over the last twenty years.
2. By how much has the temperature changed?
3. Draw a sketch graph to show what you expect the average temperature to change by over the twenty years from 2000–2020.

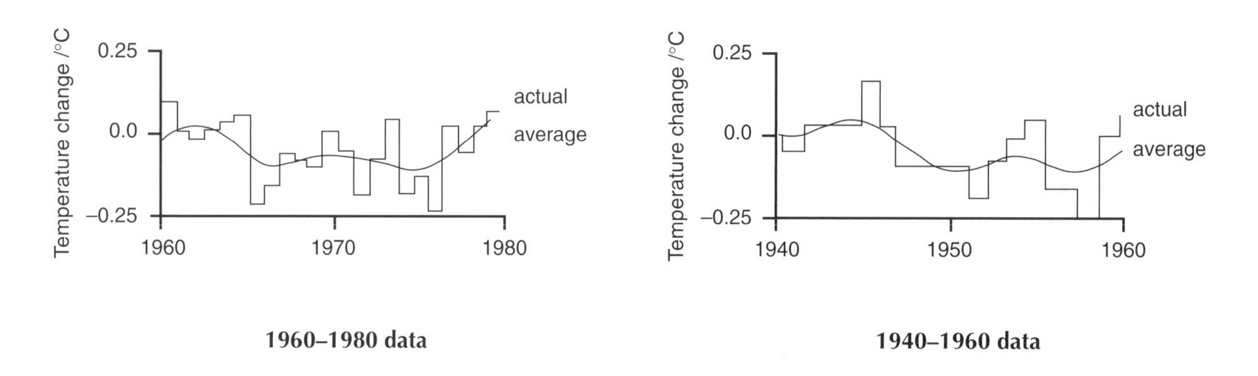

1960–1980 data

1940–1960 data

4. Compare how the temperature of the earth changed between 1960 and 1980 and between 1940 and 1960.
5. Based on the new information from question 4, draw a second sketch graph to show what you expect the temperature to change by over the twenty years, 2000–2020.

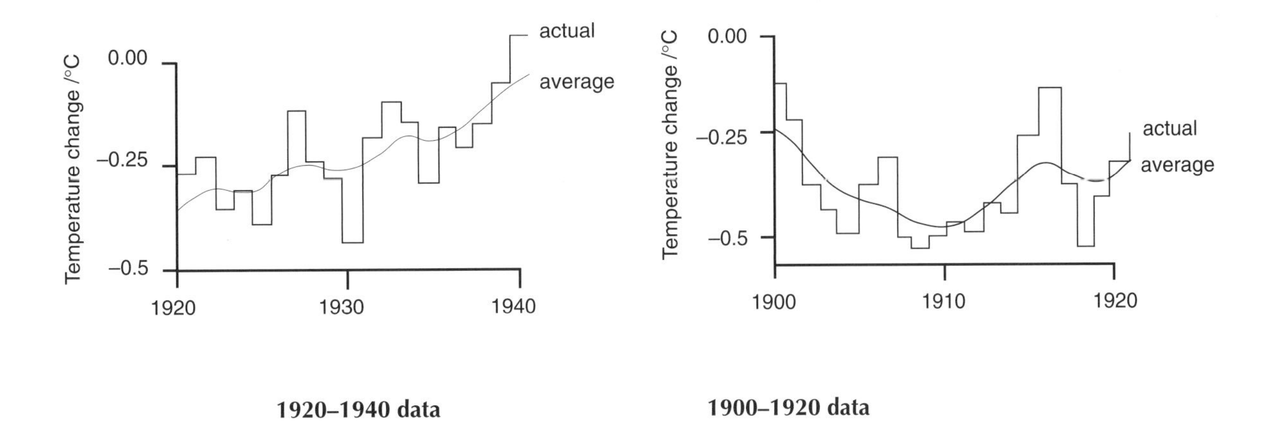

1920–1940 data **1900–1920 data**

6. Compare how the temperature of the earth changed between 1920 and 1940 and between 1900 and 1920.
7. Based on the new information from question 6, draw a third sketch graph to show what you expect the temperature to change by over the twenty years 2000–2020.
8. Has the overall temperature of the earth increased over the last century? (Use the data above to support your answer.)
9. Based on the temperature changes over the whole of the 20th century, which of your sketch graphs is most likely to be correct?
10. Do you think that you have enough evidence to support a firm conclusion to your answer to question 9?

When looking for answers scientist usually analyse data from more than one source if they can. The following graph comes from the Central England Temperature (CET) record. This record goes back to 1660 when instruments were first used to record temperatures. Annual temperatures are recorded here.

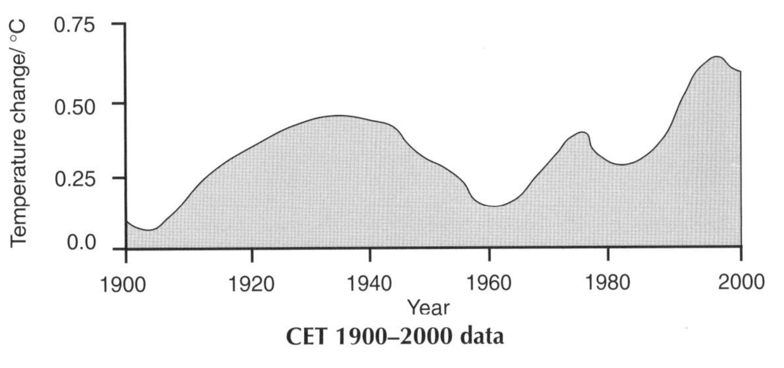

CET 1900–2000 data

11. Does the Central England Temperature Record show the same temperature pattern as the IPCC data above?
12. How does the overall temperature change compare to the answer you gave in question 8?
13. Suggest a reason why different data collected during different experiments might vary, leading to uncertain conclusions.

Looking at the data – temperature changes over several centuries

Study the graphs, from the Central England Temperature record (CET) and then answer the questions.

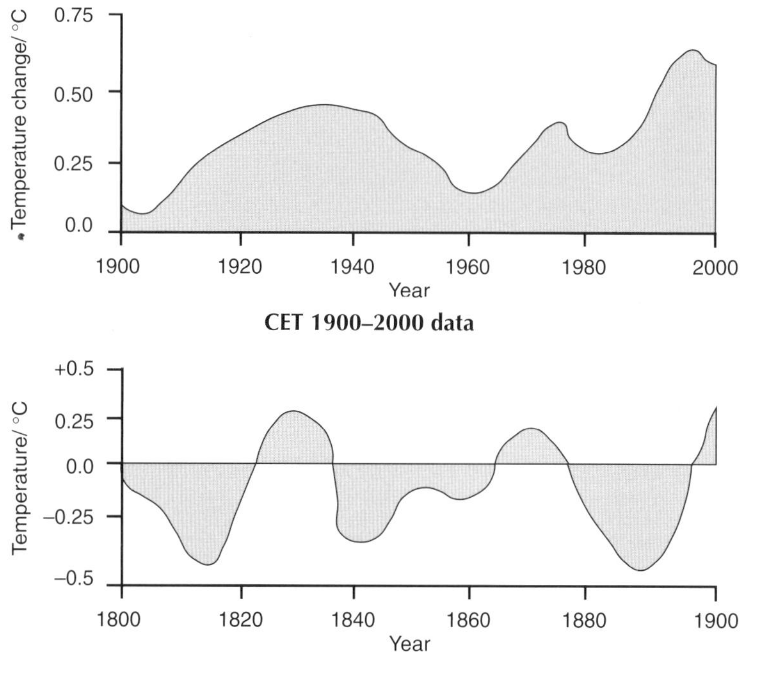

CET 1800–1900 data

1. Do you think that the temperature changes seen in the 19th century were any different to those seen in the 20th century?
2. Which century was the coldest?

You may have seen pictures of people ice skating on the river Thames. At the same time, other places in Europe were also suffering from long bitterly cold winters and cold wet summers. This cooler period is often known as the 'Little Ice Age'.

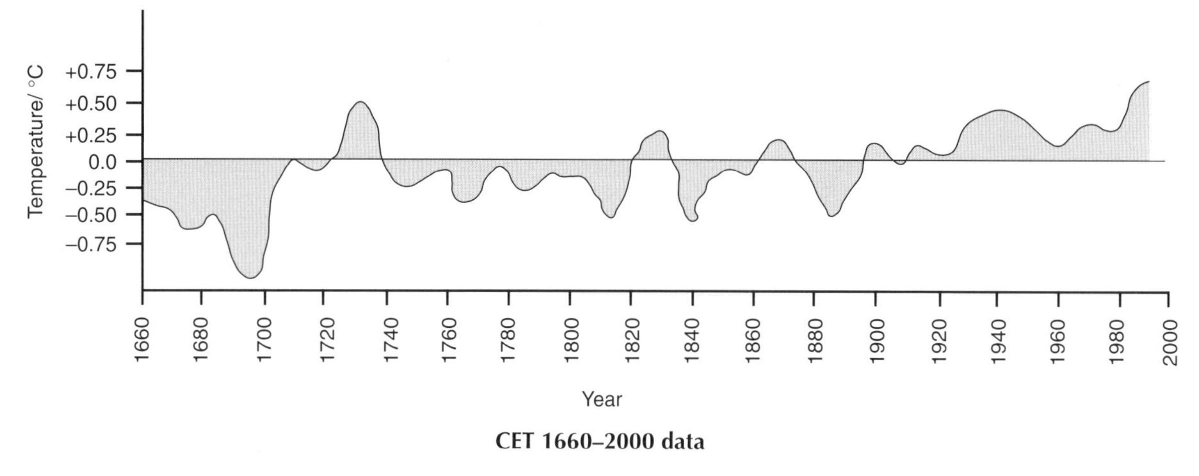

CET 1660–2000 data

3. When do you think the end of the Little Ice Age was?
4. After the Little Ice Age, how long did it take to warm up?
5. Compare the temperatures of the 1730s and 1740s with the present temperature.

6. How do these values compare to the temperature in the 1930s?

7. How do you think the temperature will vary over the next twenty years?

In groups of 3 or 4

8. Look back at the conclusions you came to about how you think the average temperature will change based on:

- the 1980–2000 data
- the 1940–1980 data
- the 1900–1940 data
- the 1660–2000 data

Try and decide if any are correct.

9. What is your overall conclusion about global warming?

10. Has the data changed your views about global warming?

If you have time, go on to the extension sheet.

Looking at the data – extension sheet

Question 1. **Looking at the data, does the evidence so far support a firm conclusion? Explain your answer.**

So far you have investigated data from two different sources. The Central England Temperature record goes back for 340 years, which is a very long experiment, but when we stop and think about the age of the earth, 340 years is a very short length of time because the Earth is millions of years old. To help support our conclusion we need to take our temperature investigation back further. To do this a new method of determining temperature (and time) must be found.

Question 2. **Can you think of anything that scientists could use to help discover what the temperature was hundreds, thousands and even millions of years ago?**

Background information

Recently leading scientists have also tried to answer the above question. Their search for complete temperature records has taken them to some places on Earth where many people do not go, such as the middle of the Atlantic and Pacific oceans, and the ice and snow of Greenland and Antarctica. The scientists have applied an idea that the 'father of geology', James Hutton had in the 1700s.

In the 1700s James Hutton from Edinburgh realised that the key to understanding the future of the planet was by investigating the past. James Hutton was the first man to study the rocks and realise that the Earth was millions of years old. He introduced the idea of geological time, and used fossils to learn about previous ages. Hutton's ideas were very controversial at the time because in 1650 Archbishop James Ussher of Ireland completed his dating of the Earth using biblical evidence.

Ussher declared that the world was created on the evening of October 22, 4004 BC, and was less than 6000 years old.

Information sheet – collecting climatic data from ice and sediment cores

Drilling into seabed sediments from sea ice presents a huge technological challenge. In 1996 drilling in the Antarctic was cancelled due to storms. In 1997 the drilling had to finish early, once again because of bad weather.

A drill rig is set up and positioned on sea ice which is about 1.7 m thick. The drill has to be lowered more than 200 m through water before reaching the seabed. It is a special design incorporating inflatable floats under the sea ice, and a submarine video camera with lights to monitor operations.

Lake sediment core from the Larsemann Hills

(Reproduced with permission from the British Antarctic Survey.)

The layers of sediments indicate the age of the sediment. Each year a fresh layer of sediments form. The composition of shells and other remains of sea life are used as a temperature indicator.

Sea bed sediments are unable to retell the whole story. They were laid down very slowly over hundreds of years, so the dating is limited.

Ice Cores

During 1997–98 scientists from British Antarctic Survey (BAS) set out to drill holes in the 4000 m high ice plateau, where the average temperatures are –50 °C. The ice cores can be used to reveal climatic data.

Rob Mulvaney and colleagues collect ice cores
(Reproduced with permission from the British Antarctic Survey.)

Each year the snow that falls on the polar cap becomes buried under new layers of snow. It becomes compressed and turns to ice. During the process bubbles of air get trapped in between the ice crystals. At a depth of 50–100 m the air is sealed and trapped from the outside world. These air bubbles contain a sample of carbon dioxide from ancient atmospheres and are easy to analyse.

It is more difficult to determine the temperature at the time the air was trapped. This is done by measuring the ratio of different isotopes ^{18}O:^{16}O and ^{2}H:^{1}H in the ice, relative to seawater. When the two ratios are low, this indicates a lower temperature. At higher temperatures more of the lighter molecules escape into the atmosphere because they are moving faster. This process also occurs when water condenses to form snow.

Collecting climatic data that is millions of years old

To investigate what the climate was like millions of years ago, geologists look to fossilised plants. By studying different types of plants they can gather climatic information, such as annual temperature range and water availability that corresponds to the time when the plant was living. This data can then be fed into computer climate models. Fossilised animals and pollen found in the same area, together with the position in the rock layer where the plants are found, are often used to age the fossilised plants.

Several different techniques are used to gather temperature information. Here are three methods that are used.

1. **Fossil flora** are compared to the nearest living plants today and the temperature is extrapolated back.

Question 1. **What do you think is the main problem with this method?**

2. **Leaves as thermometers.** The shape of the leaves, the nature of the leaf margin and the feature of the leaf cuticle can all be used to provide estimates of mean annual temperature, temperature range and water availability. The graph shows the relationship between temperature and the percentage of smooth leaves found together in an assemblage.

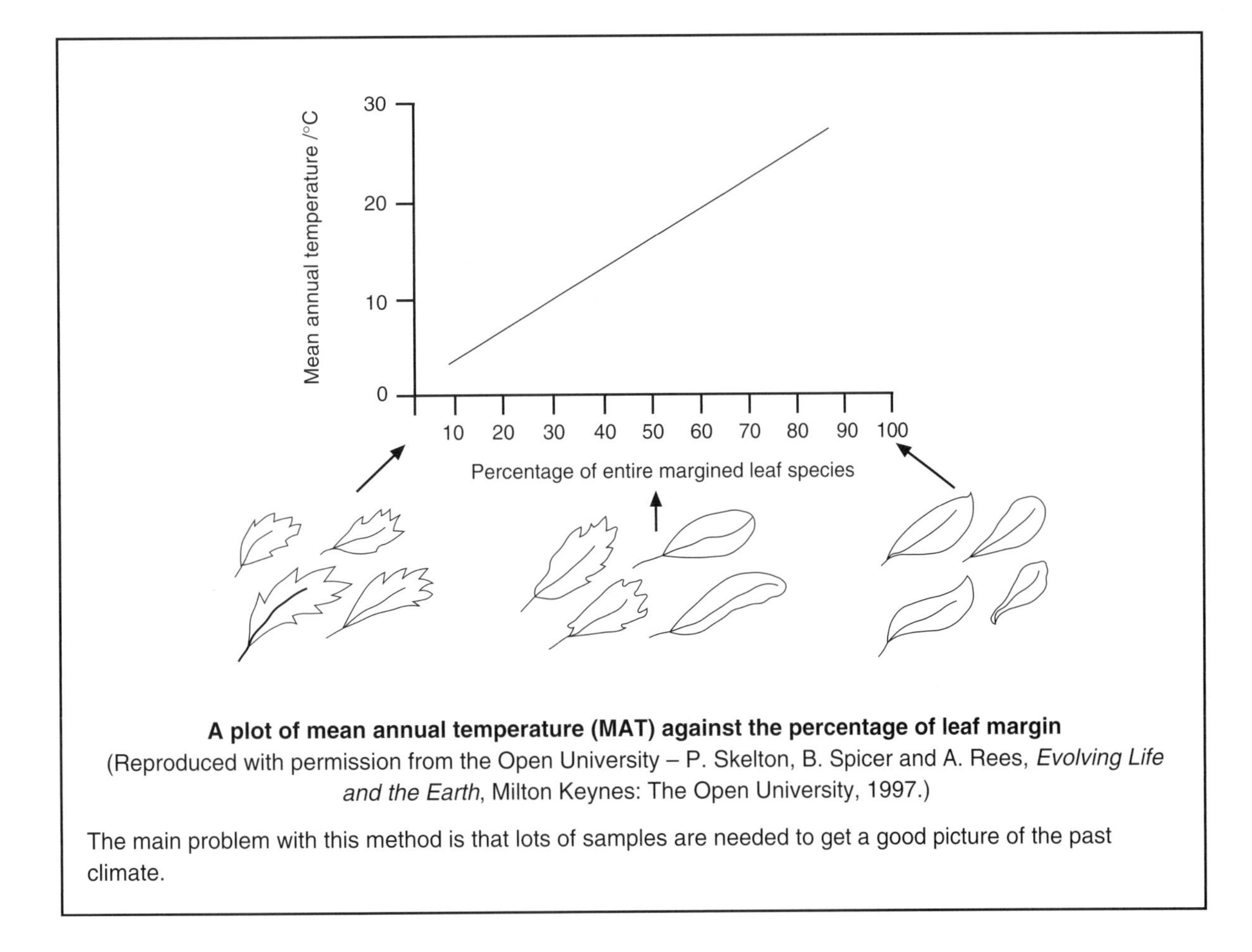

A plot of mean annual temperature (MAT) against the percentage of leaf margin
(Reproduced with permission from the Open University – P. Skelton, B. Spicer and A. Rees, *Evolving Life and the Earth*, Milton Keynes: The Open University, 1997.)

The main problem with this method is that lots of samples are needed to get a good picture of the past climate.

Question 2. **Using the graph, work out the mean annual temperatures if the following leaves were found together in an assemblage.**

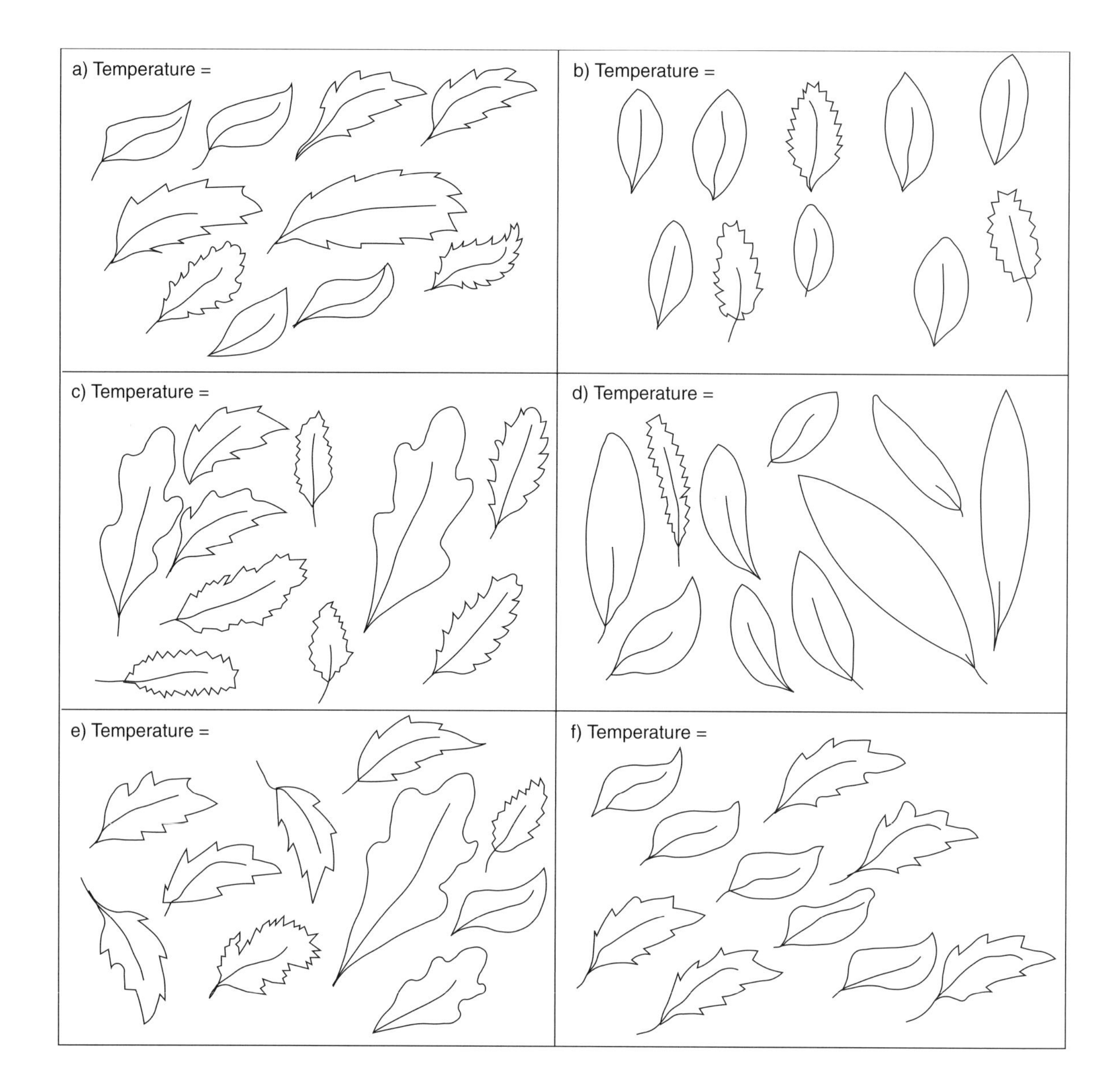

3. **Tree growth rings** in fossil wood record a pattern of annual growth over their lifetime. They can give information about temperature changes, water availability, how long the growing season was, light levels and even what insects were around!

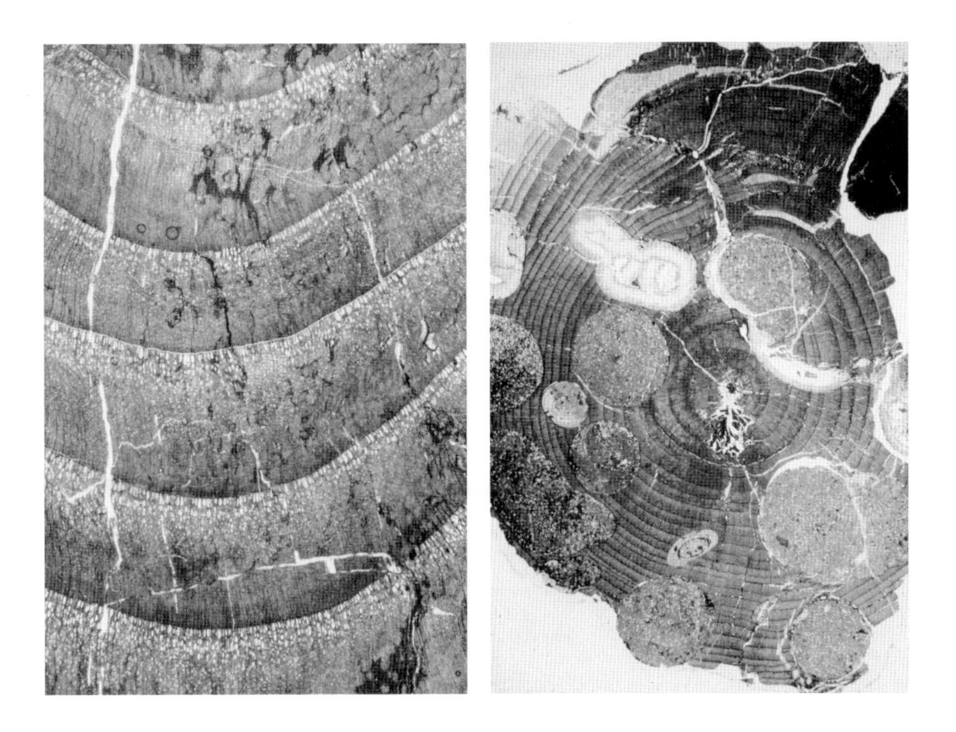

Photograph A **Photograph B**

Tree growth rings

(Reproduced with permission of Jane Francis, University of Leeds.)

Question 3

Study the two photographs and then answer the questions.

a) Which tree do you think lived in a warm wet climate? Explain your answer.

b) Which tree lived through periods of drought? Explain your answer.

c) How often did drought occur?

Extension question

How do you think the accuracy of the temperature measured by instruments today compares with early instrumental records and data collected from ocean sediments, ice cores and fossilised leaves and trees?

Looking at climatic data from the past

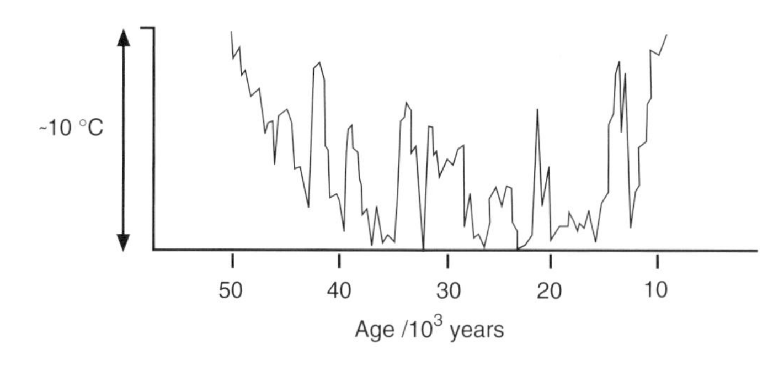

Data collected from ocean sediments of the North Atlantic Ocean – 50000 years old
(Data source: Intergovernmental Panel on Climate Change.)

Using the ocean sediment data above:

1. How has the temperature of the Earth varied over the last 10 000 years?
2. When was the last time the temperature was similar to today's temperature?
3. When was the temperature 10 °C lower than it is today?
4. Using this data, how do you think the temperature will vary over the next 10,000 years?

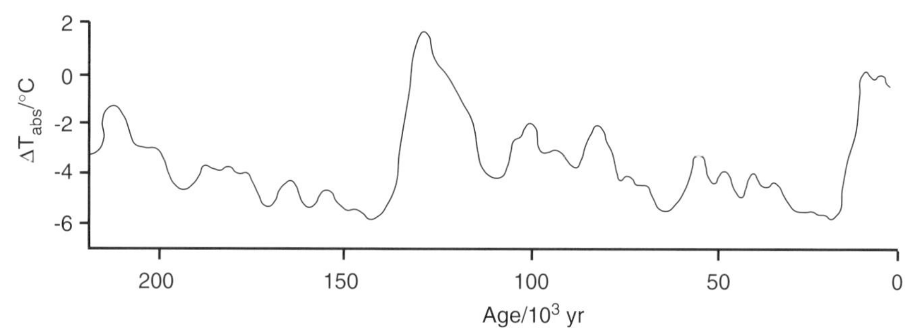

Data collected from Antarctic Ice Cores for 240000 years
(Reproduced with permission from *Nature*.)

Using the ice core data above:

5. How has the temperature of the Earth varied over the last 10,000 years?
6. When was the last time the temperature was similar to today's temperature?
7. What was the lowest recorded temperature?
8. When did this low temperature occur?
9. When was the temperature last stable for 10,000 years?
10. What happened over the next 10,000 years from the period of stability?
11. Describe what happened after that?
12. How do you think the temperature will vary over the next 10,000 years?

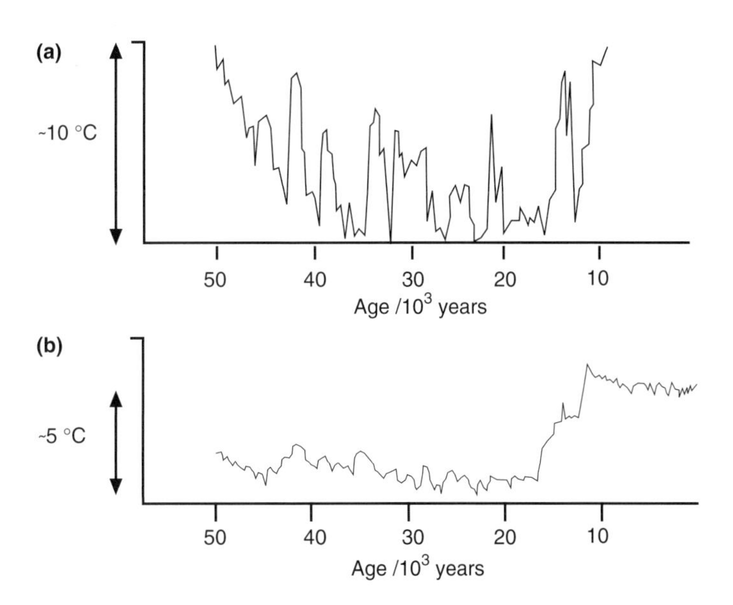

Comparing data from (a) sediments of the North Atlantic Ocean, with (b) data collected from Antarctic ice cores

(Data source: Intergovernmental Panel on Climate Change.

13. Look at the two data sets above and write down when the results agree.

14. Look at the two data sets and write down when the results disagree.

15. Do the two sets of data follow the same patterns?

16. Suggest a reason why the data sets are not the same.

17. When interpreting data and predicting future global temperatures what do you think are the main problems / questions facing scientists?

Many scientists are very concerned about the temperature of the Earth and thus many are employed to research into the past to try and understand what is happening in the present.

Data covering even longer timescales is shown on the next page.

The figure below shows how scientists think the surface temperature of the Earth has changed over 600 million years.

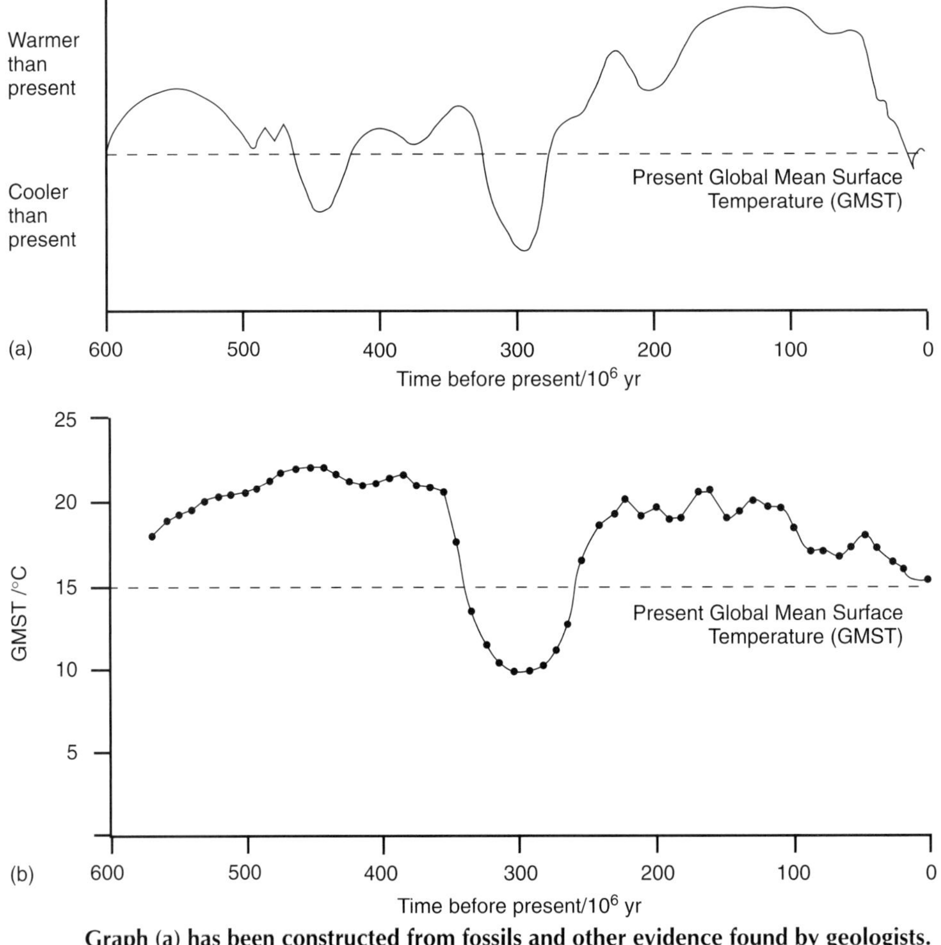

Graph (a) has been constructed from fossils and other evidence found by geologists. Graph (b) has been estimated using a computer model, which is known to scientists as the GEOCARB model

Reproduced with permission from the Open University (P. Francis and N. Dise, *Atmosphere, Earth and Life,* Milton Keynes: The Open University, 1997.)

18. Looking back over the last 2000 million years:

a) How has the temperature generally compared with today's temperature?

b) How many ice ages have there been?

c) How often does an ice age occur?

19. Does this new evidence help us to predict the future temperature of our planet?

Life as a scientist in Antarctica

Dr Jane Francis has been on geological expeditions to Antarctica. She works in the department of Earth Science at the University of Leeds. We interviewed her to find out what life was like as an expedition scientist.

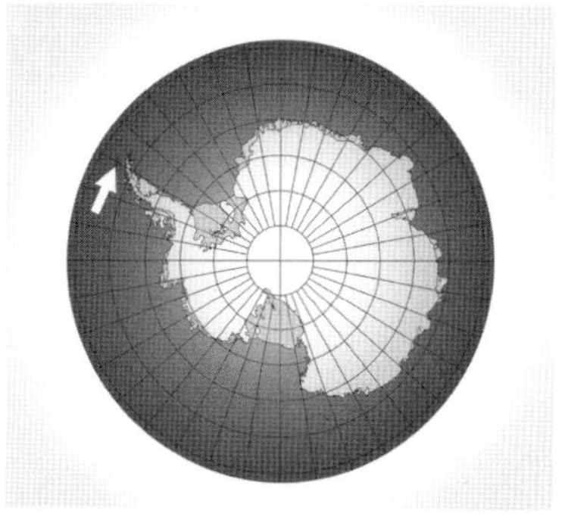

Map of the South Pole – the arrow indicates Livingstone Island
(Map reproduced by courtesy of the British Antarctic Survey.)

Q Where is Antarctica?

A Near the South Pole

Livingstone Island, Antarctic Peninsula
(Reproduced with kind permission of Jane Francis, University of Leeds.)

Q How do you get to Antarctica?

A By boat and plane. We fly from the UK to the Falkland Islands, then catch a ship to Antarctica.

Q How long does the expedition last?

A Usually 2 or 3 months during the summer, December to February. During this time it is light for most of the day and night because it is so near the South Pole.

Q Where do you stay?

A In tents as close as possible to the rocks we are working on. The tents take two people and we sleep and cook in them. They are very windproof and quite warm.

Expedition tents and storage boxes

(Reproduced with kind permission of Jane Francis, University of Leeds.)

Q Does is get very cold?

A Yes, we have to wrap up warm. I wear three pairs of socks and often five layers of clothes. Sometimes it is too cold to go outside and we have to stay in the tents (lie-ups). When it snows we read a lot of books and play games.

Generally, the weather in the summer is usually very good, clean air, blue skies, warm sun but very strong winds, so temperatures are below freezing. Also we can have blizzards and a lot of snow. We have to wear sunglasses all the time and sunscreen on our faces because it is very easy to get very sunburnt because of the ozone hole!

Q How many people in the group?

A 2–5. Each scientist has a field assistant for safety.

Q What do you eat?

A A lot of high energy dried food such as soya, dried fruit, porridge and tons of chocolate for energy. We drink a lot of tea because we get dehydrated in the cold air.

Q Where do you wash?

A We don't, its too cold! You don't smell in cold temperatures. Also, sometimes there is a shortage of water because we have to melt ice to make water and that uses up precious fuel supplies.

Q What do you do in the evening?

A It takes a lot of the evening to cook dinner on one small stove. Then we work, we have to write up the notes from the day's work.

Q Do you miss your home, while you are away?

A Yes, but the work is very interesting and Antarctica is such a beautiful and exciting place to be.

Q What work do you do?

Dr Jane Francis
(Reproduced with kind permission of Jane Francis, University of Leeds.)

A I am a geologist so I map rocks, record the rock type, I see and measure sections such as the thickness of different layers of rocks. I write lots of notes, take photos and make sketches. The rock samples and fossils that I collect are wrapped up in big boxes and taken home by ship. Months later, they arrive in the UK!

Q Where are the rocks?

A Most of Antarctica is covered by thick ice but rocks are exposed on the coast in sea cliffs, on small islands and on mountain tops which stick up through the ice caps.

Q If I wanted to be an Antarctic scientist, what qualifications would I need?

A First you must be a good scientist. In the sixth form you should study subjects such as chemistry, physics, biology, geology, geography. At university you need to study subjects such as earth sciences and geology.

Q Do you need to do any fitness training before you go on an expedition?

A You must be fit and healthy... because a lot of the work is physical and the harsh conditions mean that you often lose weight on the trip.

Q What is the first thing that you do, when you get home after a long expedition?

A Eat fresh fruit and salads, smell flowers and look at colourful things.

Q If it is cold and the food is not very good, why do you go on the expeditions?

A It's a fantastic place to be, the work is very exciting. Some scientists are involved in drilling ice cores which are then taken home in a big freezer to analyse later. I usually go looking for different types of rocks, fossils and petrified plants. Some of the things we find are millions of years old, even older than the dinosaurs. In fact about 10 years ago, we found the remains of an Antarctic dinosaur. It was in millions of pieces, many of which were buried in the sand. On the way home in the boat we had to put all the pieces together. It was the hardest jigsaw I have ever done. All the pieces looked the same!

Q I thought that dinosaurs only lived in warm places?

A Yes you are right. Our findings shows that millions of years ago the Antarctic temperature must have been much warmer, at least 18 °C warmer that it is today. Together with some of our results from other work, we believe that 70–100 million years ago Antarctica has a similar climate to New Zealand, and this climate lasted until about 40 million years ago.

Q Why was the climate so different?

A Now that's a good question. Scientists believe that the warmth of Antarctica resulted from the warm 'greenhouse' climates that affected the whole earth at that time. And these climates were probably caused by much higher levels of carbon

dioxide in the atmosphere, as volcanoes emitted this gas. However, the answer is really complicated because many other variables are involved such as ocean currents and the actual position of Antarctica may have been very different. Many scientists have put forward different theories to try and explain the climate.

Q So in the future could we see forests in Antarctica?

A Perhaps if the climate was to become very warm.

A simulation showing what Antarctica might have been like millions of years ago
(Reproduced with permission from Course S269 materials, Milton Keynes: The Open University.)

Student activity

The British Antarctic Survey (BAS) has just decided to expand its research projects in Antarctica. New scientists are needed to carry out vital research.

Your job is to write an advert to encourage people to apply for the post.

In the advert, you will need to include the required type of person required, the qualifications, and what they might be doing.

To help you research more about life in Antarctica, visit the following BAS website at **http://www.nerc-bas.ac.uk/nerc-bas.html** (accessed June 2001) and then click on information for schools and students. The Antarctic diary will tell you about life at the Rothera Station. Photographs are also included. There is a slide show collection at **http://usarc.usgs.gov/** (accessed June 2001) and more information at **http://www.antarcticanz.govt.nz/education/Pages/InfoEducation/Education.msa** (accessed June 2001).

A Global Warning

The Earth **is** getting warmer

Looking at the data – the Earth is getting warmer

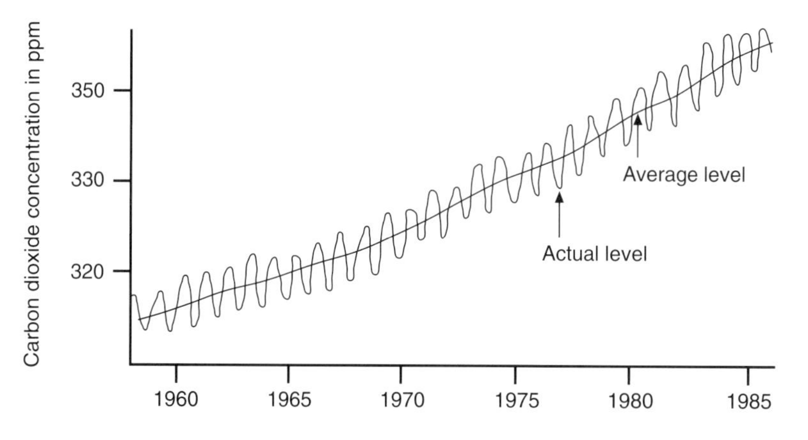

The atmospheric concentration of carbon dioxide between 1958 and 1992 at Mauna Loa, Hawaii

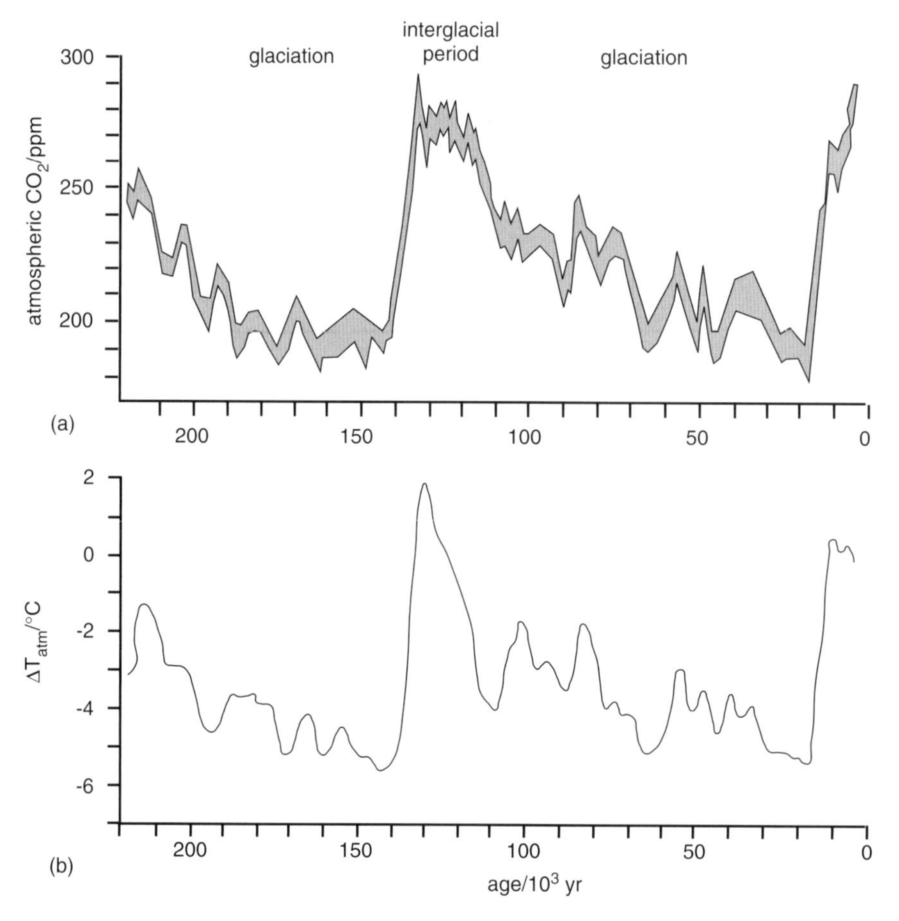

Data from Antarctica showing CO_2 levels and temperatures over the last 240,000 years
(Reproduced with permission from *Nature*.)

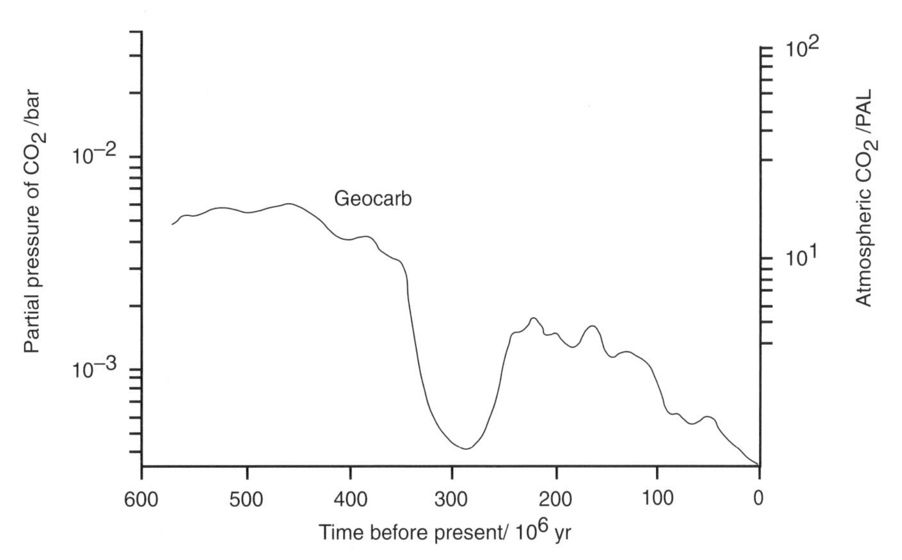

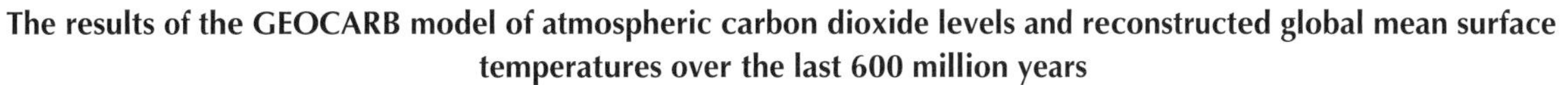

The results of the GEOCARB model of atmospheric carbon dioxide levels and reconstructed global mean surface temperatures over the last 600 million years

(Reproduced with permission from the Open University P. Francis and N. Dise, *Atmosphere, Earth and Life*, Milton Keynes: The Open University, 1997.)

Question

1. Do you think there is a link between atmospheric CO_2 levels and the average surface temperature of the Earth? Use the above data to support your answer.

Other greenhouse gases

So far we have come to the conclusion that an increase in the levels of carbon dioxide in the atmosphere accompanies a rise in global temperature. We are now going to look at some data from other greenhouse gases to see if they have the same effect.

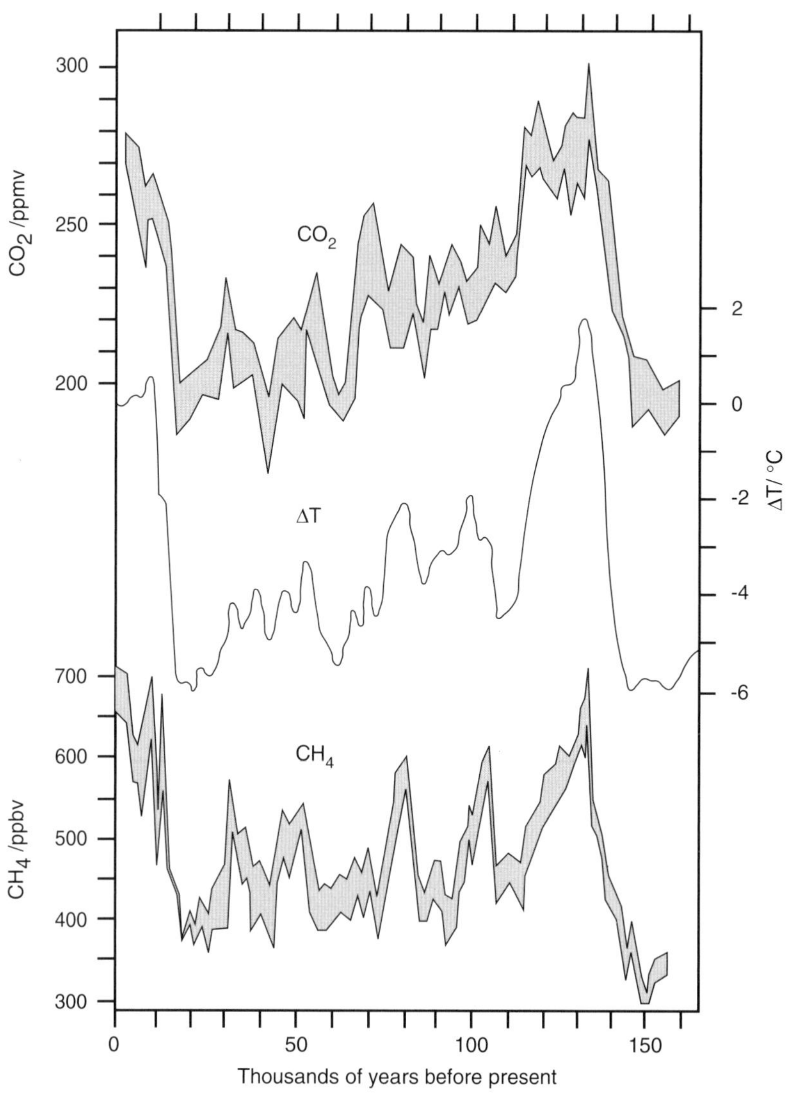

Graphs obtained from Antarctic ice core records of local atmospheric temperature and corresponding air concentrations of carbon dioxide and methane for the past 160,000 years

(Data source: Intergovernmental Panel on Climate Change.)

1. As the air temperature increases, what happens to the amount of methane in the atmosphere?
2. Using the data from the graph to support your answer, do you think the amount of methane in the atmosphere could lead to global warming?

The following graphs show estimated historical concentrations of major greenhouse gases, over the last 250 years.

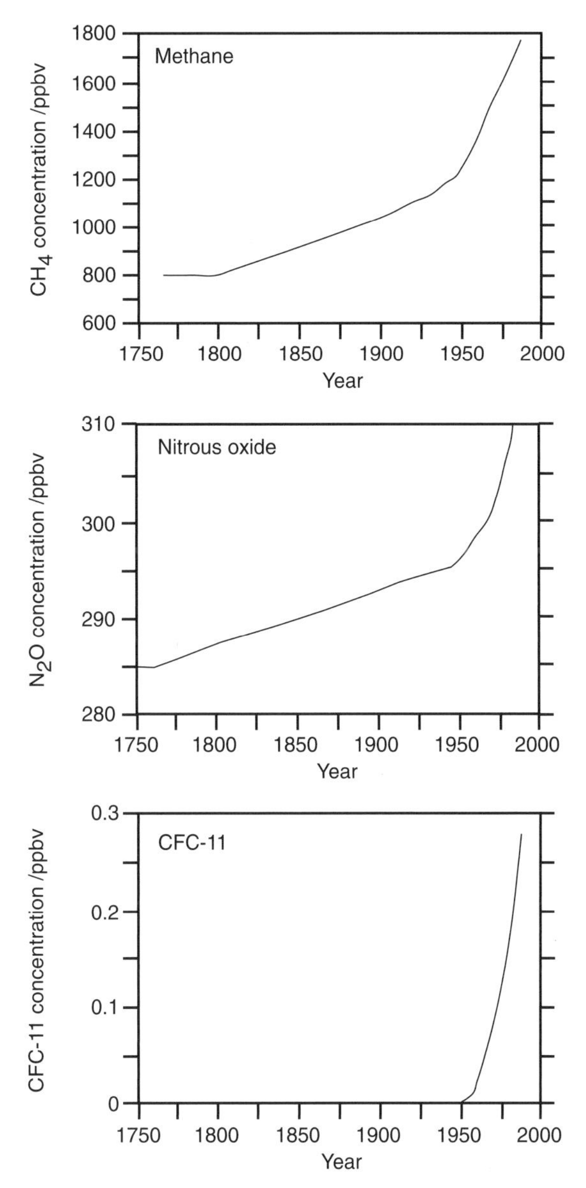

Graphs showing estimates historical concentrations of major greenhouse gases, over the last 250 years

(Data source: Intergovernmental Panel on Climate Change.)

3. In which year did the level of methane in the atmosphere first start to increase?

4. Describe what happened to the amount of methane in the atmosphere after 1950.

5. When did CFC-11 start to build up in the atmosphere?

6. The amount of nitrous oxide in the atmosphere has steadily increased during the last 250 years. When did the rate of increase suddenly change?

7. Do you think that any of the above gases could be linked to global warming? You must explain your answer.

The amounts of greenhouse gases in the atmosphere have increased dramatically over the last fifty years. This is mainly because of increased human population and activities, such as man-made chemicals and combustion.

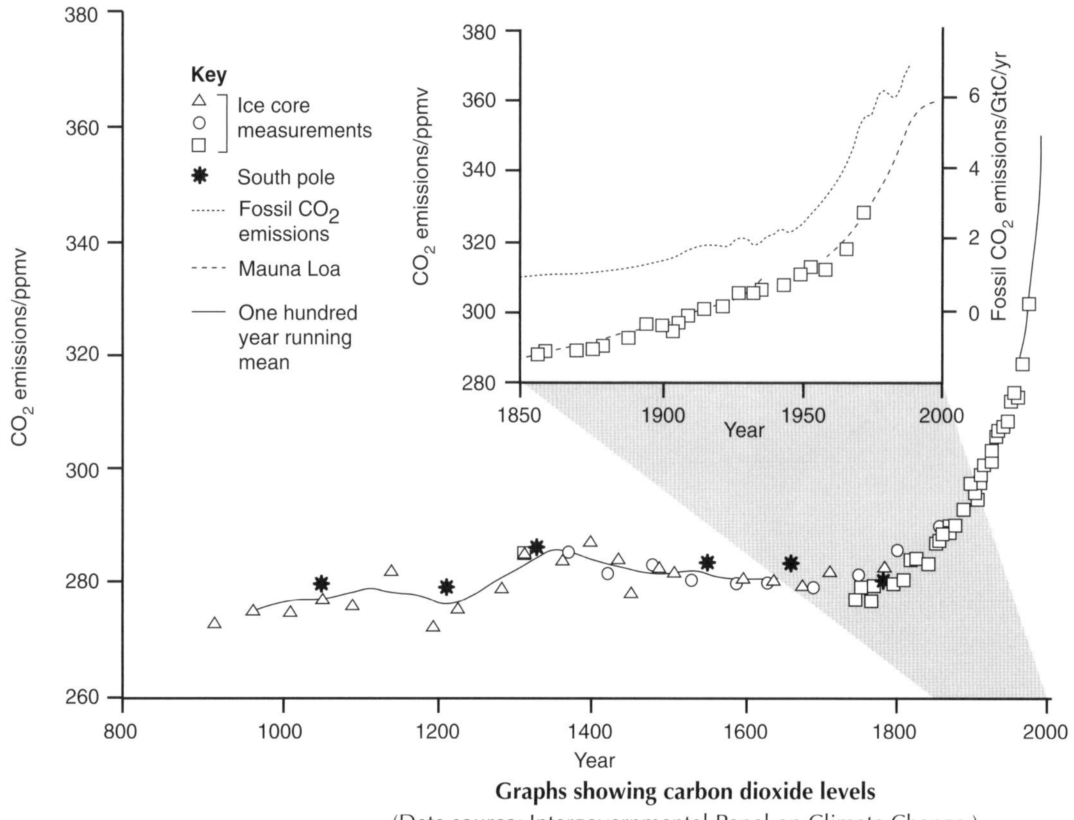

Graphs showing carbon dioxide levels

(Data source: Intergovernmental Panel on Climate Change.)

The table below lists the main man-made sources of these gases. Greenhouse gases also occur naturally.

Greenhouse gas	Sources due to human activities
Carbon dioxide	Burning of fossil fuels Deforestation
Methane	Bacteria in rice paddy fields Released from natural gas and oil wells Landfill – (getting rid of waste) Domestic animals – mostly cattle Coal mining Biomass burning
Chlorofluorocarbons (CFCs)	Refrigerants Aerosols
Nitrous oxides	Fertilisers Combustion of fuels in cars and power stations Biomass burning.

Sources of greenhouse gases

One of the big problems with greenhouse gases is that once they enter the atmosphere, it is a long time before they leave. The table below shows how long each molecule of gas will stay in the atmosphere.

	Carbon dioxide	**Methane**	**CFC-11**	**Nitrous oxide**
Atmospheric lifetime (years)	50–200	10	65	150

Atmospheric lifetime of greenhouse gases

8. Which greenhouse gas stays in the atmosphere for the shortest length of time?

9. Which greenhouse gas is the hardest to get rid of?

10. You work for the local council and have been invited to give a speech on global warming and the greenhouse effect. Members of the audience will be local residents and business people. Prepare notes for the speech.

Your speech should include the following points;

- What the greenhouse effect is;
- The likely causes of recent global warming (over the last fifty years);
- Your plans to help reduce global warming by local actions;
- The problems you expect to find when carrying out these plans; and
- The time-scale needed to carry out these plans.

You may like to include a visual aid to use during your speech.

Theories about global warming

Future climates – activity page

Take part in the biggest experiment ever!

Help to model the climate of the 21st century. Further details at **www.climate-dynamics.rl.ac.uk/~hansen/casino21.html** (accessed June 2001)
Design a leaflet to advertise the experiment. On it you must explain what a computer model is and why they are used.

Each year the number of cars on the roads in Britain increases and therefore so does the total amount of CO_2 cars produce. Do you think we should try not to use our cars so much? Design a questionnaire to find out what people in your school think, how people travel to school and if they are willing to change. What should future transport be like? You could write a letter to your local MP telling him or her the results of your project.

Find out what scientists think will happen in the future. Start your search at **www.gcrio.org/ipcc/qa/06.html** (accessed June 2001) or **www.meto.gov.uk/sec5/CR_div/Amin/ghg.html** (accessed June 2001).

Write a report of what you find, include graphs or relevant quotations.

Trees use carbon dioxide from the air. People burn the forests so they can use the land for agriculture. The fire produces carbon dioxide, and there are then fewer trees to use it up.
Do you think this should be allowed? Find out if woods have vanished from your area. Start by comparing old ordnance survey maps with up to date ones. You could visit the local library.

Is carbon dioxide the only greenhouse gas offender or is the amount of other greenhouse gases in the atmosphere increasing?

Work through the **Other greenhouse gases** sheet to answer the question.

RS•C

Mario Molina puts ozone on the political agenda

Teachers' notes

Objectives

- To illustrate how scientific theories can influence politics and manufacturing industry.
- To interpret real ozone data.
- To understand that, over time, the composition of the atmosphere has changed and that human influence is responsible for some of the changes.
- To know what chlorofluorocarbons (CFCs) are and their uses.

Outline

The student material is divided into three different sections:

- An information sheet on Mario Molina
- Understanding ozone
- The CFC- ozone story

Two versions of the material have been included:
Version 1 is aimed at the more able 14–16 year old student, providing plenty of opportunity for project work including searching for data on the Internet, interpreting articles and analysing data in order to make an informed decision on environmental issues.
Version 2 is a simpler version, focusing on how ozone protects the earth from UV radiation, what would happen if there was a hole in the ozone layer and what all the fuss about CFCs is really about. A timeline activity is also provided to put the material in context.

Teaching topics

This selection of activities is suitable for 14–16 year olds and could be included when teaching about the properties, reactions and uses of the halogens or about the atmosphere. It could also be used when teaching about health, safety and risk.

Background information

From Molina's initial discovery in December 1973 right up to the present day, CFCs have been discussed by scientists, politicians, research scientists, industrialists, environmental groups and ordinary people. The subject has been, at times, controversial and in the early years the scientific data was limited, the chemistry of the stratosphere was not well understood and some pressure groups tried to say ozone depletion was due to natural causes and not man-made chemicals. As more scientific evidence was collected, showing that ozone depletion was due to man-made chemicals, worldwide governments worked together to ban CFC production. Eventually industry (in developed countries) agreed to stop making the chemicals, just as some developing countries were starting to set up CFC production plants and so a separate agreement had to be made with them.

Rowland and Molina were faced with a real problem of ethics. Should they tell the world and try and stop ozone production or should they just get on with the next piece of science? This work could be used to present this type of dilemma, and question the responsibility of scientists and the scientific world. There are many newspaper articles which could be used to start discussion, such as 'Pressure on the aerosol business' and 'First moves towards a CFC free Britain', both of which have been included at the end of these teaching notes. Both of these articles show how different groups, such as industry and Friends of the Earth, responded to the threat of ozone destruction. Both articles include some background information to the CFC-ozone problem.

Ozone in the troposphere – health risks

Ozone is a poisonous gas. The World Health organisation recommends a maximum hourly dose of 80 ppb. Many countries give ozone alerts when ozone levels are high. During such an alert, people, especially children and the elderly, are advised to stay inside. The table lists observed symptoms at different ozone levels.

Ozone dosage (hourly levels ppb)	Symptoms
50	Headaches
150	Eye irritation
270	Coughs
290	Chest discomfort

Table 1 Ozone dosage

Ozone chemistry of the stratosphere

Ozone is produced continually in the upper stratosphere where UV radiation from the sun dissociates molecular oxygen to form atomic oxygen.

$O_2 + hv \rightarrow O + O$

$O + O_2 \rightarrow O_3$

The reaction occurs very rapidly in the stratosphere over the tropics, where solar radiation is most intense. Circulation in the stratosphere constantly moves ozone away from the tropics towards the poles.

Ozone is destroyed when it constantly absorbs UV light that would otherwise reach the Earth's surface.

$O_3 + hv \rightarrow O_2 + O$

There is no net ozone depletion because the process produces atomic oxygen that reacts with molecular oxygen to produce another ozone molecule.

Ozone is continually being destroyed through reactions with naturally occurring radicals of Cl, N, H or O atoms. The ozone hole problem started to occur when the concentrations of chlorine radicals in the stratosphere started to increase as a result of man-made products. The natural cycle of ozone production and destruction was put out of balance, leading to an overall ozone deficit.

Mechanism for CFC-ozone destruction:
First CFCs break down to form chlorine (Cl) radicals.

$CFCl_3(g) + hv \rightarrow CFCl_2(g) + Cl(g)$

$CF_2Cl_2(g) + hv \rightarrow CF_2Cl(g) + Cl(g)$

RS•C

The chlorine radicals then react with ozone in a chain reaction.

$O_3(g) + Cl(g) + hv \rightarrow O_2(g) + ClO(g)$

$ClO(g) + O(g) \rightarrow O_2(g) + Cl(g)$

The overall effect on ozone is:

$O_3(g) + O(g) \rightarrow 2O_2(g)$

Sometimes the ClO produced may react with nitrogen compounds but more chlorine radicals are then produced:

$ClO(g) + NO_2(g) \rightarrow ClONO_2(g)$

$ClONO_2(g) + HCl(s) \rightarrow Cl_2(g) + HNO_3(s)$

$Cl_2(g) + hv \rightarrow 2Cl(g)$

The chlorine radicals then react with ozone as follows:

$2Cl(g) + 2O_3(g) \rightarrow 2ClO(g) + 2O_2(g)$

$2ClO(g) + M \rightarrow Cl_2O_2(g) + M$ where M is a third body

$Cl_2O_2(g) + hv \rightarrow ClO_2(g) + Cl(g)$

$ClO_2(g) + M \rightarrow Cl(g) + O_2(g) + M$

The overall effect on ozone is:

$2O_3(g) \rightarrow 3O_2(g)$

The dramatic seasonal ozone depletion comes at a time of year when there are no oxygen atoms present. In the stratosphere, a stream of air known as the polar vortex circles Antarctica in winter. Air trapped within this vortex becomes extremely cold during the polar night. Temperatures drop low enough to form clouds. The polar stratospheric clouds provide surfaces for chlorine producing reactions (as shown above). By spring the stage is set for chlorine to chew up ozone as the sun rises and ends the long Antarctic polar night. Sunlight splits the molecular chlorine into chlorine atoms that attack ozone, forming molecular oxygen and ClO. The ClO forms a dimer, which in turn, is photolysed to chlorine atoms, which attack more ozone, forming a hole. The hole disappears when the polar vortex finally breaks down after the spring sun warms the air over the Antarctic. Air then sweeps in from lower altitudes, bringing nitrogen oxides that tie up the active chlorine and ozone that fills the hole.

Teaching tips

This topic presents several opportunities for group discussions on topics such as scientific ethics, how scientists communicate their work and the responsibility scientists and consumers have to protect the environment.

When introducing this work, it is extremely important to stress that the CFC-ozone story continues today. Scientists monitor the amount of ozone in the stratosphere daily, and it is this data that will be interpreted during the lesson.

The information sheet on Mario Molina can be used to set the scene either by recounting the story to the class or by getting the students to read it for themselves.

The student sheet 'Understanding ozone' introduces the students to ozone.

The CFC-ozone story student sheet offers a structured approach to telling the story and interpreting ozone data.

The timeline

- This provides a way of telling the story and it sets a context for students to relate to in terms of other things that were happening at the same time. Students should be encouraged to add to the timeline as they research the topic further.
- Making the timeline may not be appropriate for all students. If you feel that the task is not demanding enough for the class, give them a ready made timeline and ask them to discuss in groups the different ways the scientists communicated with the world and the response that the world made.

Interpreting the data

If possible the students look up and download their own data from the Internet. The advantage of the students going to the websites themselves should reinforce the fact that there are many scientists monitoring ozone levels on a daily basis. The topic they are learning about is undergoing scientific investigation all the time.

For those who do not have web access, ozone data obtained by the British Antarctic Survey has been included for 1999-2000.

Resources

- Glue & scissors
- Internet access
- Student worksheets:
  – The timeline
  – Mario Molina (1943–) information sheet
  – Mario Molina version 1 – understanding ozone and
  – Mario Molina version 1 – the CFC-ozone story
  or Mario Molina version 2

Timing

Approximately 30 minutes if given the outline or 60 minutes for groups making up their own outline for the timeline.

One or two lessons or homework for the work with Mario Molina.

Opportunities for ICT

Using the Internet to obtain up to date information.

RS•C

Newspaper Articles

Pressure on the aerosol business

by Derek Harris

Britain's aerosol industry is squaring up to resurgence of the ozone controversy, one result of which could mean large capital spending on new equipment and some company closures with job losses.

It could create a particular problem for ICI as principal supplier in Britain of the aerosol propellants called chlorofluorocarbons. These could be outlawed because it is claimed they thin the ozone layer in the stratosphere.

The ozone layer protects the earth from the sun's ultra-violet radiation. An increase in radiation is likely to cause a greater incidence of skin cancer in white people.

Although evidence on ozone depletion has yet to emerge, Sweden is banning most aerosol sprays from January next year. In the United States, Oregon has brought in a shop ban on many aerosols - while allowing hairdressers, for instance, to buy and use aerosol hairsprays. After that questionable start federal agencies have moved in with a ban timetable that will stop the manufacture after October 15 of 'non-essential' aerosols using as propellants the chlorofluorocarbons, otherwise known as CFCs.

That means in the United States that a third of the goods bought in aerosol packages, such as hair perfume sprays and deodorants, will have to switch to a different propellant not implicated in the ozone controversy, the rest having already ceased using CFCs.

American manufacturers have switched largely to using hydrocarbons like butane or propane as propellants. But in Europe about 70% of aerosols at present use CFCs as propellants, while in Britain the proportion is probably slightly higher.

This is why United Kingdom aerosol fillers and the CFCs' producers are anxious how far and how quickly the EEC will follow in American footsteps. There has been much pressure in Holland, for a ban on CFC aerosols and it is on the cards that the EEC will decide later this year to start a review of the situation.

Studies on the effect of CFCs are already being carried out in this country and West Germany, adding to the research already being done in the United States.

In terms of collected evidence the ozone controversy is at a stage where at any rate doubts can validly be raised about the continued use of CFCs. But the evidence is largely the rest of work on mathematical models, which in itself has produced questions of validity.

Some counter theories are being advanced which, if proved right, could turn what looked like an ozone disaster into at least a manageable problem and possibly barely a problem at all. But it is likely to be several years before there is conclusive evidence.

That leaves the manufacturers of aerosol-packaged products and the can fillers (not all manufacturers fill their own cans) weighing the question of when to spend their money on change and, indeed, what change.

Aerosol packaged goods are a £250m a year industry at retail sales values. Last year 532.5 million cans were filled with products ranging from insecticides and medical products to paints, foods and artificial snow as well as the toiletry products, which make up half the total sector.

Hair sprays are far the most popular aerosol product, accounting for some 30% of total aerosol production.

Companies like Unilever's Gibbs, Beechams, Reckitt & Colman and the Wellcome Foundation are among the manufacturers involved, but there are also contract fillers of which Aerosols International, part of Cadbury Schweppes, is by far the largest.

The options open to the industry are limited. One answer is as quickly as possible to drop the use of CFCs except for the specialist applications for which there is no substitute, such as in medical products like the bronchodilators used by asthmatics.

That would almost certainly mean a switch to the use of the hydrocarbons, which are already used in Britain as elsewhere, particularly in products, which have a water base such as starches and polishes. Hydrocarbons are cheaper - CFCs being three times the price - but they are also flammable.

At one time some of the smelly molecules - usually sulfur derivatives - in butane/propane mixtures made them unsuitable for applications like toiletries, but much purer hydrocarbons from this point of view are now available.

However it poses problems for those making up a propellant "cocktail" for a particular product because of the desirability of countering the flammability. There are some solubility problems compared with

CFCs. But it is the flammability, which poses the biggest cost problem in that if a can-filling factory is not equipped for hydrocarbons large changes are necessary.

Special storage facilities are needed together with other increased safety arrangements in the factory and also in the supply chain after the product has left the factory gate.

For most manufacturers the cost of factory installations alone is likely to run from between £100,000 and £250,000. It is this sort of cost which smaller fillers may not be able to meet. There are around 120 fillers altogether in the United Kingdom, eight being major manufacturers and 20 particularly small.

Some in the industry believe enough of the smaller establishments would be driven out of business to put at risk at least 1,500 out of the 10,000 jobs in the industry.

Nobody believes it would be acceptable to consumers to go back in applications like hair sprays to the old finger-operated pumps that pre-dated the aerosol packages. The use of carbon dioxide or nitrogen with no flammability problem but producing a coarse and too variable a spray, offers no scope although the possibility of a combination with hydrocarbons is being looked at.

ICI, which has a big stake in CFCs not only in producing for the aerosol market, but also in such applications as refrigerants, has been looking at alternative CFCs.

One possibility is to produce a less stable CFC, which would be broken down during its journey to the stratosphere, thus rendering it harmless to the ozone.

RS•C

First moves toward CFC free Britain

Jonathan Porritt

By the end of this year, 90 per cent of aerosols on sale in the UK will be CFC-free. Since aerosols have, until now, accounted for more than 60 per cent of CFC use in this country, eliminating them from this particular industrial sector was obviously the single most important thing Britain could do to help protect the ozone layer.

CFCs - chlorofluorocarbons - are contained in the propellant that carries liquid drops from the nozzle of an aerosol can (and also used in refrigerators and air-conditioning units). Once seen as the perfect chemical - odourless, non-flammable and chemically inert - CFCs are so stable that they can hang around the atmosphere for more than 100 years.

However, they also destroy the ozone layer that protects the earth from about 99 per cent of ultra-violet radiation by releasing chlorine as their molecules break down.

Friends of the Earth's campaign to persuade the aerosol manufacturers to phase out CFCs was launched in 1986, and was over by 1988.

First we published our pamphlet, The Aerosol Connection, a detailed list of aerosols, which were not using CFCs.

This was coupled with as much publicity as we could generate at the time to encourage consumers to find out which aerosols they should be buying.

When this "softly-softly" approach failed to elicit anything other than vaguely hostile rebuffs from the aerosol manufacturers, we felt it necessary to prepare an outright boycott of the best-selling CFC-based products in the UK.

The aerosol industry's decision to get out of CFCs by the end of 1989 was taken just three days before the boycott campaign was launched.

In the light of subsequent events, this was obviously a sound decision. But it was actually based on the fear of consumers turning against all aerosols, not just CFC-based aerosols, rather than on any rational assessment of the scientific position.

Consumer awareness is often a somewhat rudimentary weapon, but the industry accurately read the signs of what was happening. Once the Prince of Wales declared that he had banned all aerosols from his household, they knew they were fighting a losing battle.

As a result, the Government found itself in the enviable position of being able to claim international credit for meeting the Montreal Protocol's original target of a 50 per cent reduction in CFC consumption a full 10 years ahead of the target date.

It was this breakthrough, which has allowed them to campaign so actively for an 85 per cent reduction.

But it is important to realize that the Government had nothing to do with this achievement. Until 1987, the Government was lobbying, primarily at ICI's behest, for a freeze on CFC production or, at best, a mere 20 per cent reduction within the Montreal Protocol. Its much-vaunted "voluntary approach" was all but worthless, in that it meant little more than leaving it to voluntary organizations such as Friends of the Earth and the Consumers' Association.

And there are other cautionary postscripts. In the first place, the Government's skilful handling of its propaganda, portraying itself as "the saviour of the ozone layer" has persuaded many people that the problem has been comprehensively dealt with, and that Friends of the Earth should now direct its attention elsewhere.

As it happens, this is far from true. The US Environmental Protection Agency presented some stark predictions to the recent conference in Helsinki on the Montreal Protocol, indicating that ozone levels are unlikely to stabilize at their 1985 levels until around the year 2070, even if we could completely eliminate all CFCs and other ozone-depleting chemicals by the end of 2000.

Second, there is no evidence to indicate that the overall sales of aerosols were affected in any lasting way. Production of aerosols in 1990 is still expected to be more than 800 million units.

Friends of the Earth therefore takes the position that its success is relative. If we have encouraged individuals to set out on the long green road to genuine sustainability, through more environmentally-sensitive lifestyles, we are well pleased.

But if this surge of consumer power amounts to no more than a panic response to the threat of increased skin cancer, then it would be wrong to wax too lyrical about its long-term environmental benefits.

RS•C

Answers

Mario Molina puts the atmosphere and ozone on the political agenda – version 1

A. Understanding ozone

1. Sunbathing

	True	False
A sun tan is healthy		Yes
A tan will protect you from the sun		Yes
You can get burnt on a cloudy day	Yes	
You can get burnt if you are in water	Yes	
With sunscreen to protect me, I can sunbathe for much longer.		Yes

B. The CFC-ozone story

2. Carbon, fluorine, chlorine, covalent bonding.

Timeline questions 1,2,3

Level of response marking could be used here.

Evidence questions

1. From this data only approx. 1970.
2. Yes, the graph shows that the October level of ozone is still going down.
3. The amount of ozone depleting chemicals in the atmosphere should peak around 2000, this means that the hole in the ozone layer should stop getting bigger. It will be about 2045 before the amount of ozone depleting chemicals reach the level they were at before the hole was first identified.
4. If the Montreal Protocol and later amendments had not taken place then the amount of ozone depleting chemicals in the stratosphere would have increased from 2 ppb in 1980 to 20 ppb in 2055. This would have destroyed even more ozone, leading to devastating effects on plant and marine life as well as increased cases of skin cancer and cataracts. Instead it is predicted that by 2055 the amount of ozone depleting chemicals will be back to the levels in 1980 and the hole in the ozone well on the way to recovery.
5. December
6. October
7. Up to 100 Dobson units.
8. Figure 2 shows ozone levels at about 300 Dobson units in Octobers before1997, whereas present October levels are at about 100 Dobson units. A drop of 200 Dobson units!
9. End of December and the beginning of January.
10. August / September

RS•C

11. As the temperature increases so does the level of ozone in the stratosphere.

12. See the notes above about the polar vortex (page 46).

Optional questions

13. Camborne in Cornwall and Lerwick in Shetland.

14. Annual rate of change in ozone levels is recorded at –0.32% at Camborne and –0.3% at Lerwick.

15. Total Ozone Mapping Spectrometer.

16 & 17 The data is available it just needs to be found!

The story continues

Teachers will need to use their professional judgement in assessing questions 18–20.

Mario Molina – version 2

1. From left to right, toxic, oxidising agent, irritant.
2. Accept general answers for the first question such as irritates the throat or eyes, toxic if too much is breathed in etc.
3. Life needs to be protected from the UV radiation in the sun.

	True	**False**
A sun tan is healthy		Yes
A tan will protect you from the sun		Yes
You can get burnt on a cloudy day	Yes	
You can get burnt if you are in water	Yes	
With sunscreen to protect me, I can sunbathe for much longer.		Yes

4. Carbon, fluorine and chlorine.
5. Aerosol propellant, foams, air conditioners, refrigerants.
6. Producing products that the consumer wanted such as hair spray, deodorants etc, different types of foams for furnishings.
7. He thought that they might destroy stratospheric ozone. This would mean that harmful UV rays would reach the Earths' surface.
8. They thought that the experiments would take too long, if they were right immediate action would be required.
9. The ozone level as been showing a steady decrease since about 1970. Before then, the level was constant at about 300 Dobson units.
10. Around 1970
11. December
12. October
13. Up to 100 Dobson units.
14. Figure 2 shows ozone levels at about 300 Dobson units in Octobers before1997, whereas present October levels are at about 100 Dobson units. A drop of 200 Dobson units!

RS•C

15. See the notes above about the polar vortex (page 43).

Teachers will need to use their professional judgement in assessing questions 16–18.

The timeline

The Ozone Story		What else is happening?
	1970	
	1971	
	1972	
	1973	
	1974	
	1975	
	1976	
	1977	
	1978	
	1979	

1980

1981

1982

1983

1984

1985

1986

1987

1988

1989

1990

1991

1992

1993

1994

1995

1996

1997

1998

1999

2000

Making the timeline

In the table you will find a number of dates and events, listed in column 1. Cut them out, sort them into the correct order and then stick them on the left hand side of your timeline. If you find out any extra information, add this on to your timeline. In column 2, there are other dates; cut them out and stick them on the right hand side of your timeline. You can also add on other important events.

Column 1	Column 2
1971 It was thought that exhaust gases from a future fleet of supersonic aircraft might damage the ozone.	I was born
6/1975 Oregon becomes the first state to ban CFCs in aerosol sprays.	1980 John Lennon killed in New York.
9/1987 The Montreal Protocol is signed, calling for eventual worldwide CFC reduction of 50% by 1999.	1979 Mrs Thatcher becomes prime minister.
1995 The production of CFCs is banned in developing countries.	I started secondary school.
6/1991 Mount Pinatubo erupts in the Philippines, speeding up the conversion of chlorine already in the stratosphere into forms that destroy ozone.	I started primary school.
Worldwide deadline for zero production of CFCs 2010	1997 Tony Blair became Prime Minister.
1995 Mario Molina shares the Nobel Prize in Chemistry with Sherwood Rowland and Paul Crutzen for his work on CFC and ozone.	1997 Diana Princess of Wales was killed in a car crash.
8 / 1985 NASA's satellite photos confirm the existence of an ozone hole over Antarctica.	1989 The Berlin Wall came down.
12/1973 Molina & Rowland discover that CFCs can destroy ozone in the stratosphere.	1982 The Falklands war.
6/1974 The CFC-ozone theory is published in Nature.	1991 The Gulf War started.
9/1974 The CFC-ozone theory is discussed in public at the American Chemical Society. 10/1978 CFCs used in aerosols are banned in the United States.	1976 The first 'pop video'. Queen and Bohemian Rhapsody.
12/1988 Preliminary findings of an ozone hole over the Arctic are discussed at a scientific conference in Colorado	1990 Nelson Mandela freed from prison
1990 Montreal Protocol strengthened in London. A complete ban on CFCs by the end of the century.	1984 Band aid and feed the world. The first 'pop' fund raising concert with Bob Geldolf.
1992 Copenhagen amendment to the Montreal Protocol. A complete ban on the production and use of CFCs by 1996.	1977 The release of Star Wars.
1985 The Vienna convention called for additional research and international exchange of ozone depletion information.	1988 The first GCSEs were taken.
3/1988 Dupont changes mind and announces that it will stop making things with CFCs.	1981 The opening of the Musical 'Cats' by Andrew Lloyd Weber. Now the longest running musical.

1988 An international meeting confirms that chlorine compounds were the cause of the ozone loss over the Antarctic.	**You may wish to add more things to this side of the timeline.**
1991 Rate of ozone depletion increasing over heavily populated areas.	
1986 Solomon's Antarctic Expedition provided strong evidence for manufactured chlorine compounds destroying ozone.	
1994 Many companies had now stopped making CFC based products.	
1993 Rowland speaks to the public to try and dispel rumours that the cause of ozone destruction was from natural causes.	
1987 Solomon's 2nd Antarctic expedition confirms manufactured chlorine compounds destroy the ozone.	
1986 Neil Harris examines ozone records taken at Arosa Switzerland. His results confirm the same ozone depletion.	
1997 Montreal amendment to the original Montreal Protocol.	
1995 Vienna amendment to the Montreal Protocol.	

Mario Molina (1943–) information sheet

Mario Molina
(Picture reproduced courtesy of Nobel Foundation.)

As a boy, Mario Molina was strongly influenced by his aunt, a chemist in the sugar industry, who later became a teacher. She used to encourage him to carry out chemistry experiments at home in a converted bathroom. From a young age Mario's ambition was to be a research scientist, even though it was not a trendy job for a young Mexican.

Why investigate CFCs and the atmosphere?

Molina went to university and studied chemistry at degree level. He then took a research degree (a PhD) in 1972, at the University of California, Berkeley. Molina then went to Irvine to work with a man called Sherwood Rowland, who had recently heard that the British scientist James Lovelock had discovered some of the refrigerant trichlorofluoromethane (called CFC-11) in the atmosphere of the Northern and Southern hemisphere. He was curious to find out more, and wanted to know the answer to a simple question, 'what happens to CFCs in the environment and were there any consequences?'

Rowland managed to persuade his sponsors to fund the project and Molina started investigating CFCs in October 1973, even though his knowledge of atmospheric chemistry was limited.

Molina got to work, carrying out calculations and he soon started to build up a very worrying picture of the atmosphere. If he was right, it was not good news; if he was wrong he would look stupid. What should he do next?

Molina's theory

CFC's were so inert that there was nothing for them to react with in the atmosphere. So air currents carried them up into the stratosphere, where energy from ultraviolet (UV) radiation would break off a chlorine atom, called a radical. The radical would then start a chain reaction with ozone that would eventually destroy the ozone layer. At the then current CFC atmospheric release rate, Molina calculated that between 7 and 13%

of the ozone would soon be destroyed. This could cause problems since it was known that the ozone layer protected the Earth from harmful UV radiation.

Action

Towards the end of December 1973, Molina discussed his theory with Rowland. At first they both tried to find a mistake in the calculations, but they could not. So just after Christmas 1973, the two scientists went to visit some atmospheric chemists for a second opinion. It was known that nitrogen oxides could destroy ozone and other investigations to do with the release of hydrogen chloride from volcanoes and the ammonium perchlorate fuel planned for the space shuttle were being carried out. No one had yet investigated CFCs; the rough estimates suggested they were perhaps a factor of 100 more significant than the fuel from the space shuttle as a potential source of stratospheric chlorine.

Telling the world

After Molina made his initial discovery, he knew that if he were right, then the Earth would be in serious trouble. CFC molecules can stay in the atmosphere for about 130 years. As a scientist he felt that he had a responsibility to tell the world, and to do something about the ever-growing CFC industry.

Even though there was no experimental evidence, Rowland and Molina published the CFC–ozone theory in the scientific journal *Nature* in June 1974.

Response

Initially there was no response from the scientific world. Concerned that their voices may go unheard, Rowland and Molina discussed their theory for the first time in public, at the American Chemical Society meeting in Atlantic City, in September 1974. This time 'possible ozone depletion' hit the headlines, Molina and Rowland recommended a complete ban on the future release of CFCs to the environment. This triggered an enormous response from governments, industry, the public and environmental groups such as Greenpeace and has subsequently led to measures to reduce and eventually eliminate their use.

Mario Molina puts the atmosphere and ozone on the political agenda

(Version 1)

A. Understanding ozone

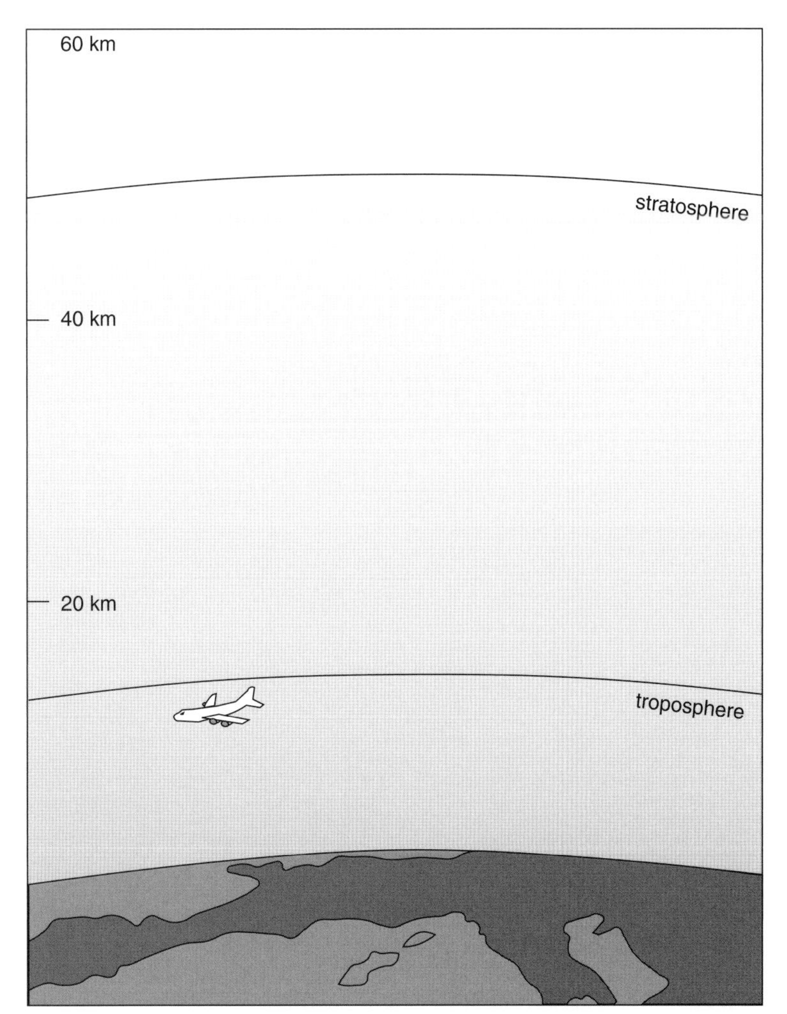

Diagram of our atmosphere

(Reproduced with permission from The Ozone Layer, UNEP/GEMS Environment Library No. 2, 1987, UNEP, Nairobi, Kenya.)

The chemical formula for ozone is O_3. The molecule contains three oxygen atoms. Ozone is quite harmful and is described by the following hazard symbols.

Ozone forms at ground level when pollutants such as nitrogen oxides and unburnt hydrocarbons react in sunlight.

Ozone in the stratosphere absorbs and protects the earth from harmful UV radiation, and is known as the ozone layer. Exposure to too much UV radiation leads to skin cancer and damages plants.

Question 1 How much do you know about sunbathing? Complete the table by ticking the correct box.

	True	**False**
A sun tan is healthy		
A tan will protect you from the sun		
You can get burnt on a cloudy day		
You can get burnt if you are in water		
With sunscreen to protect me, I can sunbathe for much longer.		

Sunburn facts

Ozone forms naturally in the upper atmosphere. Oxygen from lower levels rises into the stratosphere where it absorbs the sun's energy in the shorter wavelengths of ultraviolet radiation. This separates the two atoms in the molecules. These free atoms then combine with other oxygen molecules to form ozone.

$$O_2 \xrightarrow{\text{short wavelength UV}} 2O$$

$$O_2 + O \longrightarrow O_3$$

$$O_3 \xrightarrow{\text{long wavelength UV}} O_2 + O$$

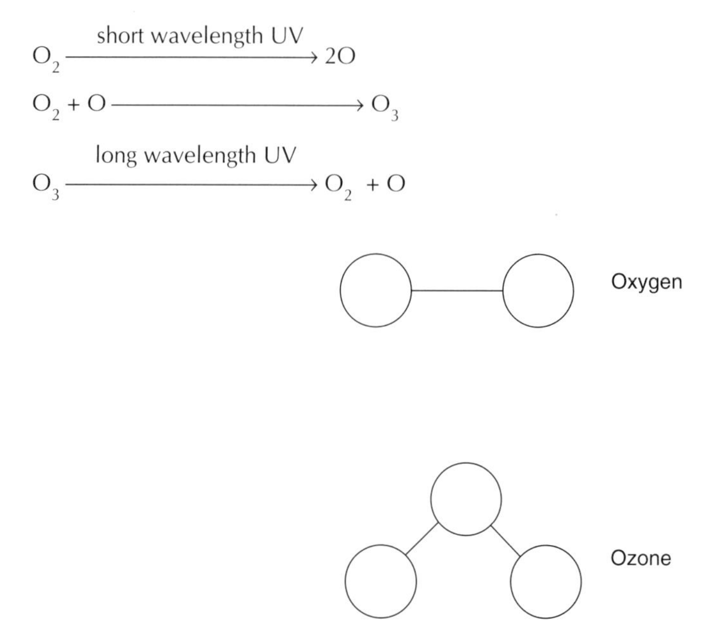

Oxygen and ozone molecules

The amount of ozone does not build-up endlessly since the energy from longer wavelengths of ultraviolet radiation breaks down the ozone molecules to produce oxygen. This process is greatly assisted by the presence in the high atmosphere of substances, such as oxides of nitrogen, which act as catalysts.

There is a very fine balance between the constant ozone production and destruction. Almost all the ultraviolet light from the sun which reaches the atmosphere is absorbed by the ozone layer. If this balance is upset, and too much ozone is destroyed, then UV light would damage plant and marine life and crop production, as well as causing skin cancer and cataracts in humans.

B. The CFC-ozone story

CFCs

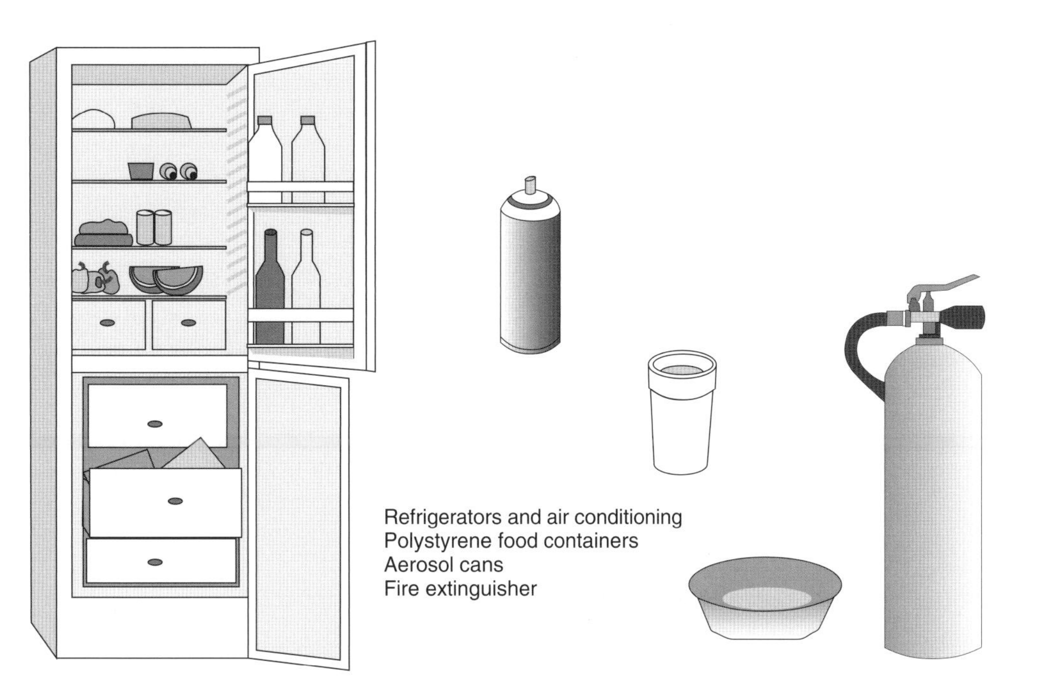

The booming CFC industry of the early 1970s

$CFCl_3$ is a CFC used in air conditioners and refrigerators.

Question 1 Name the elements in a CFC molecule.

The molecules are very stable.

What type of bonding do they have?

The problem

After carrying out some calculations in 1973, Mario Molina, a research scientist, believed that CFCs could destroy the ozone layer in the stratosphere, and the Earth would no longer be protected from the harmful UV radiation.

The CFC-ozone story

The CFC-ozone story can be told by using a timeline. Your teacher will either supply you with a ready made timeline or give you a worksheet, so that you can make your own.

Timeline questions

1. Imagine that it is 1975, and you are working for a company such as Du Pont, that produces CFCs. What is your reaction when the State of Oregon bans the use of CFCs in aerosols?
2. Why do you think Du Pont announced that it would stop CFC production in 1988?
3. Do you agree with Du Pont's 1988 decision?

It is important to realise that the story is not yet over. Every day scientists record the ozone level in the atmosphere, alternative chemicals to CFCs are being researched, and it will be a long time before the hole in the ozone layer is gone.

Looking at the evidence

When Rowland and Molina first told the world their CFC-ozone theory in 1974, it was all based on theory, with no experimental evidence. Laboratory experiments later confirmed and modified the model. The amount of ozone in the stratosphere has been closely monitored at the Halley (since 1956), Rothera (since 1997) and Vernadsky / Faraday (since 1957) stations in Antarctica.

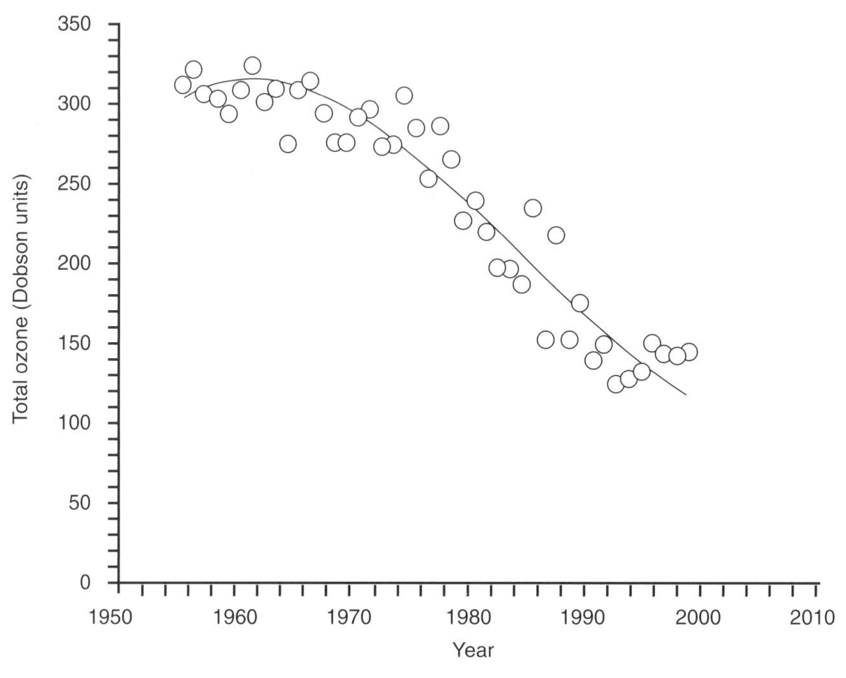

The mean October ozone levels recorded at the Halley station
(Reproduced with kind permission of J Shanklin, British Antarctic Survey (BAS).)

Evidence Questions

1. In which year did the ozone levels start to decrease?
2. From the data, do you think the ozone hole is still getting bigger?

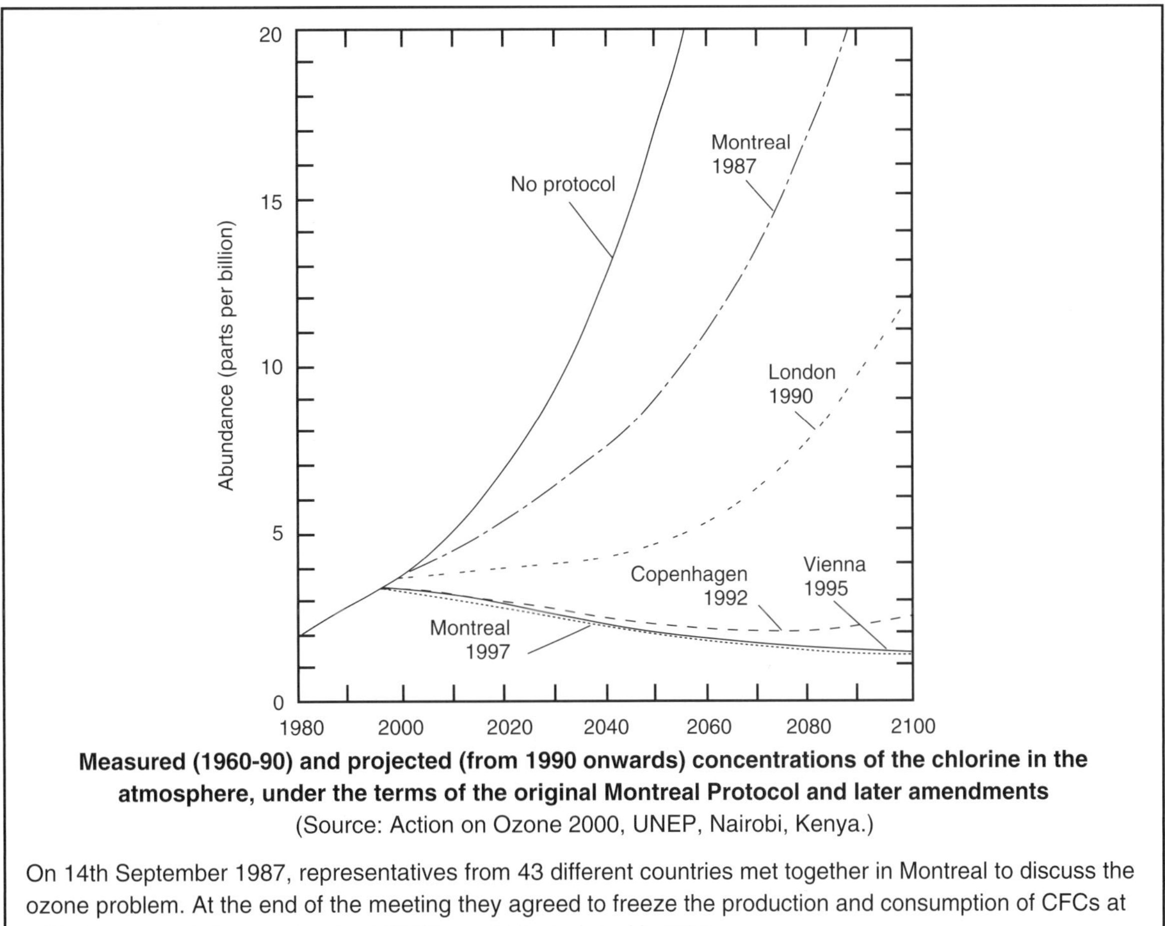

Measured (1960-90) and projected (from 1990 onwards) concentrations of the chlorine in the atmosphere, under the terms of the original Montreal Protocol and later amendments
(Source: Action on Ozone 2000, UNEP, Nairobi, Kenya.)

On 14th September 1987, representatives from 43 different countries met together in Montreal to discuss the ozone problem. At the end of the meeting they agreed to freeze the production and consumption of CFCs at 1986 levels. By 1999, the levels of CFCs would be reduced by 50%.

3. Study the figure above showing the concentration of chlorine in the atmosphere and use the data presented here to either support or reject your answer to question 2.

4. In your own words explain why the Montreal Protocol and the later Amendments were a major breakthrough for the protection of the environment.

At the Antarctic Stations, constant monitoring of ozone levels has revealed that ozone levels naturally fluctuate throughout the year.

Either visit the British Antarctic Survey (BAS) website **http://www.nerc-bas.ac.uk/public/icd/jds/ozone** (accessed June 2001), look up the ozone data or look at the data your teacher has given you. Then answer the following questions:

5. In which month of the year are the ozone levels highest?

6. In which month of the year are the ozone levels lowest?

7. How much do the ozone levels fluctuate on a daily basis?

8. Compare the present ozone level with the levels recorded before 1977.

9. In which month of the year is the temperature in the stratosphere highest?

10. In which month of the year is the temperature in the stratosphere lowest?

11. Can you find a relationship between ozone levels and temperature in the stratosphere?

12. Suggest a reason why ozone levels fluctuate.

Optional – you will need access to the Internet to work through this section

The Meteorological Office makes regular measurements of ozone at two sites in the United Kingdom. They use TOMS to provide accurate, detailed information. Visit their website at **http://www.metoffice.gov.uk/research/stratosphere/ozone/index.html** (accessed June 2001) (there is a direct link from the BAS site) to find out:

13. Where are the Metereological Office stations?

14. What long-term trends are seen at each station?

15. What does TOMS stand for?

16. Find a satellite picture showing the ozone hole.

17. Find the ozone level over the UK today.

To answer questions 16&17, you may need to surf other websites. A good place to start is at the Centre for Atmospheric Science, Cambridge University, with the 'The Ozone Hole Tour'. **http://www.atm.ch.cam.ac.uk/tour/index.html** (accessed June 2001)

The story continues...

CFCs and Ozone still makes the papers...even with all the data from scientific research, it is still a controversial subject. In developing countries economic reasons have meant that these chemicals are still being used, and even in the developed world there is still controversy. Read the following newspaper extract.

Greens see red / Overseas news
World summary

Sydney: The environmental group Greenpeace has asked a court to order the Olympics Co-ordinating Authority to stop styling the 2000 Olympics as the Green Games because it says that an ozone-depleting chemical will be used in the cooling system at one of the venues

30 March 1999, 'The Times', p.15

18. Do you support the views of Greenpeace? Visit the Greenpeace website at **http://www.greenpeace.org** (accessed June 2001) to find out more.

19. Research and find out about methods of cooling *ie* refrigerants and air conditioning systems that do not use CFCs.

20. Write a word-processed letter to the Sydney 2000 Olympic Games Committee, either supporting Greenpeace or supporting the Olympics Co-ordinating Authority, on the subject of 'Green Games'. You should include scientific / technological evidence to back up your opinion.

Mario Molina puts the atmosphere and ozone on the political agenda

(Version 2)

Ozone has three oxygen atoms. Ozone is a strong smelling, pale blue gas, which is poisonous to humans. Ozone is described by the following hazard symbols.

1.

Use the following words to label the hazard symbols.

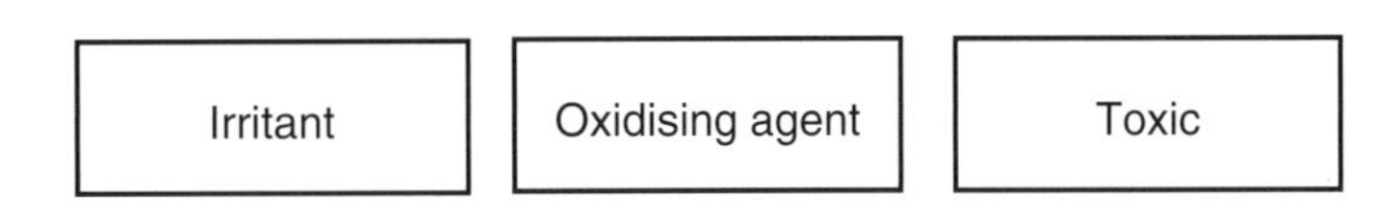

2. How do you think you would feel if you were exposed to ozone?

____

____

____

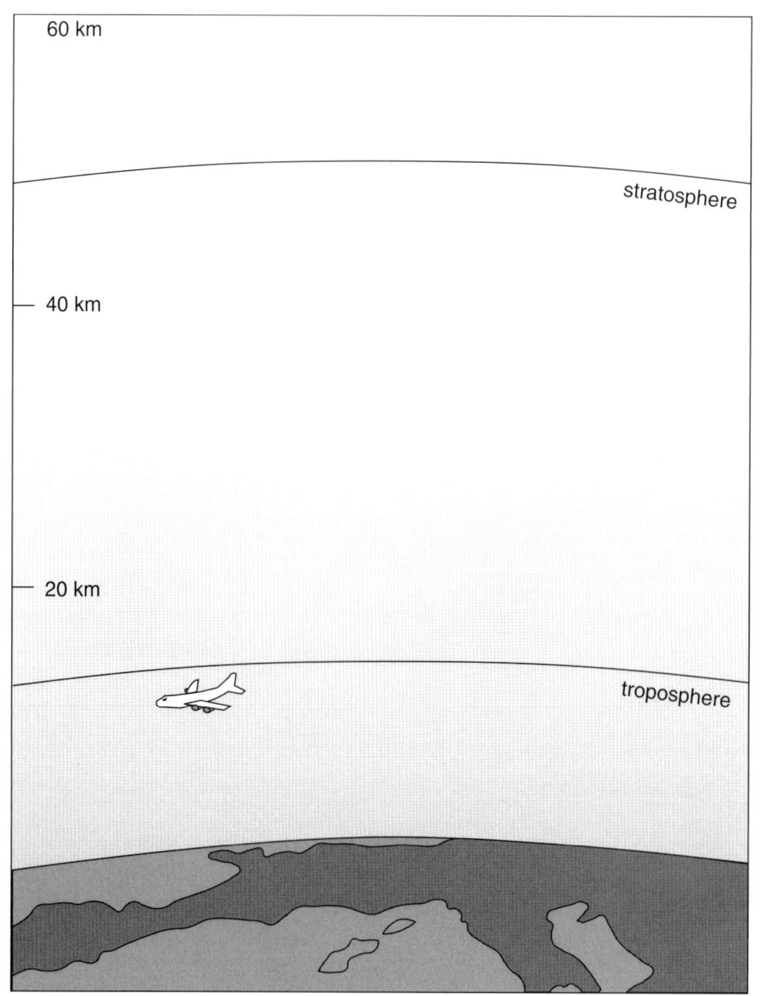

A diagram of our atmosphere

(Reproduced with permission from The Ozone Layer, UNEP/GEMS Environment Library No. 2, 1987, UNEP, Nairobi, Kenya.)

The diagram of the atmosphere shows that ozone exists in the troposphere and the stratosphere.

Ozone in the stratosphere absorbs and protects the Earth from harmful UV radiation. Exposure to too much UV radiation leads to skin cancer and damages plant and marine life.

3. Why do you think it is important to look after the ozone layer?

____

Ozone protects us from the sun

How much do you know about sunbathing? Complete the table by ticking the correct box.

	True	**False**
A sun tan is healthy		
A tan will protect you from the sun		
You can get burnt on a cloudy day		
You can get burnt if you are in water		
With sunscreen to protect me, I can sunbathe for much longer.		

After carrying out some calculations in 1973, Mario Molina, the research scientist, believed that CFCs could destroy the ozone layer in the stratosphere, and the Earth would no longer be protected from the harmful UV radiation.

$CFCl_3$ is a CFC used in air conditioners and refrigerators.

4. Name the elements in a CFC molecule.____________________________

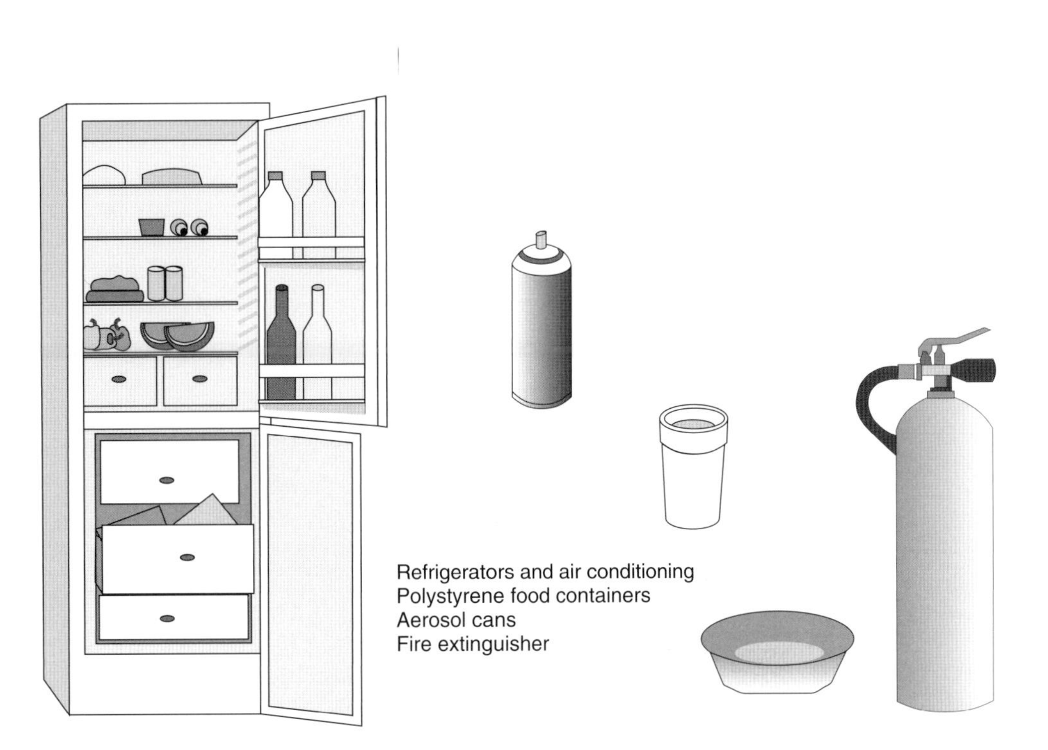

The booming CFC industry of the early 1970s

5. Give four uses of CFCs.

____

____

6. Why do you think the CFC industry was doing so well in the early 1970s?

____

____

After Molina made his initial discovery, he knew that, if he was right, then the Earth would be in serious trouble. As a scientist he felt that he had a responsibility to tell the world, and to do something about the ever-growing CFC industry.

7. What did Mario Molina believe CFCs would do to the Earth?

____

____

You now could make a timeline, which tells the CFC-ozone story so far. Your teacher will give you the instructions, which are on a separate sheet. It is important to realise that the story is not yet over. Every day scientists record the ozone level and alternative chemicals to CFCs are being researched because it will be a long time before the ozone hole is mended.

Looking at the evidence

The amount of ozone in the stratosphere has been closely monitored since 1956, at the Halley, Rothera and Vernadsky / Faraday stations in Antarctica. Scientists have shown that the amount of chlorine in the stratosphere has rapidly increased since the late 1970s.

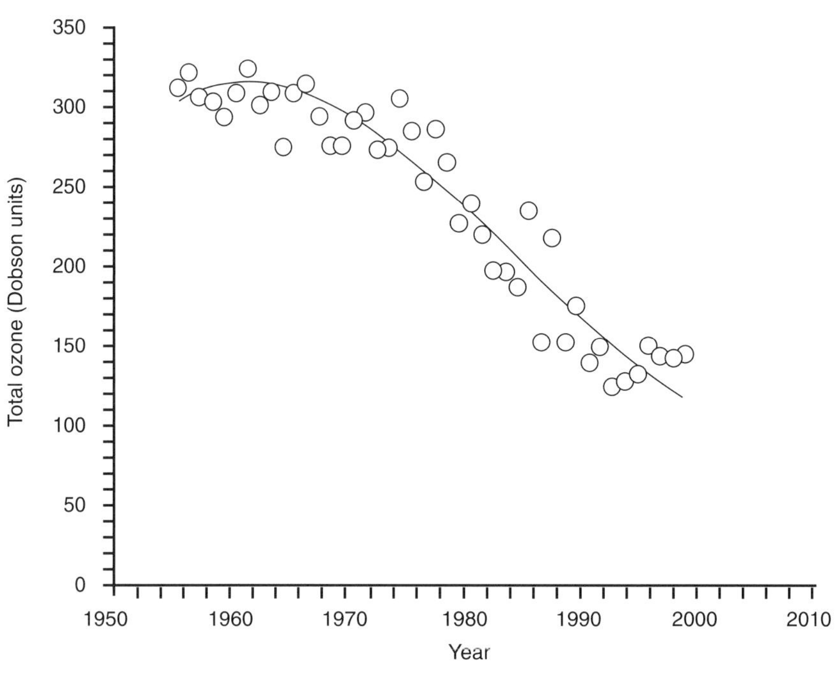

The mean October ozone levels recorded at the Halley station
(Reproduced with permission from J. Shanklin, British Antarctic Survey.)

9. In your own words describe what the graph tells us.

____

____

10. In which year did the ozone levels start to decrease?

____

Constant monitoring in the Antarctic has revealed that ozone levels naturally fluctuate throughout the year.

Real ozone data

11. In which month of the year are the ozone levels highest? ________________

12 In which month of the year are the ozone levels lowest? ________________

13. How much do the ozone levels fluctuate on a daily basis? ________________

____

____

14. Compare the present ozone level with the level recorded before 1977.

____

____

15. Suggest a reason why ozone levels fluctuate.

____

____

____

____

The story continues...

CFCs and ozone still makes the papers...even with all the data from scientific research, it is still a controversial subject. In developing countries economic reasons have meant that these chemicals are still being used, and even in the developing world there is still controversy. Read the following newspaper extract.

> **Greens see red /Overseas news**
> *World summary*
>
> Sydney: The environmental group Greenpeace has asked a court to order the Olympics Co-ordinating Authority to stop styling the 2000 Olympics as the Green Games because it says that an ozone-depleting chemical will be used in the cooling system at one of the venues
>
> *30 March 1999, The Times, p.15*

Answer the following questions

16. Do you support the views of Greenpeace?

17. Research and find out about methods of cooling *ie* refrigerants and air conditioning systems that do not use CFCs.

18. Write a wordprocessed letter to the Sydney 2000 Olympic Games Committee, either supporting Greenpeace or supporting the Olympics Co-ordinating Authority, on the subject of 'Green Games'. You should include scientific / technological evidence to backup your opinion.

RS•C

Bibliography

J. E. Andrews, P. Brimblecombe, T. D. Jickells, P. S. Liss, *An introduction to environmental chemistry,* Oxford: Blackwell Science Ltd., 1996.

G. Best, *Environmental pollution studies,* Liverpool: Liverpool University Press, 1999.

J.E.Francis, *Terra Antarctica* 1996, **3(2)**, 135-140.

J.E.Francis, *Terra Antarctica Reports,* 1999, **3**, 43-52.